Chinese Studies in History

SPRING-SUMMER 1987/VOL. XX, NO. 3–4

Recent Studies of the
Boxer Movement

GUEST EDITOR: David D. Buck
The University of Wisconsin—Milwaukee

Available in the United Kingdom and Europe from M. E. Sharpe,
Publishers, 3 Henrietta Street, London WC2E 8LU.

Published simultaneously as Vol. XX, No. 3–4 of *Chinese Studies in
History*.

Library of Congress Cataloging-in-Publication Data

Recent Chinese studies of the Boxer Movement.

  "Most of the papers in this collection were originally presented at
an international conference on the Boxer Movement held in Jinan, Shandong, in November 1980 to mark the eightieth anniversary of the
event"—Introd.
  Bibliography: p.
  1. China—History—Boxer Rebellion, 1899–1901.
I. Buck, David D., 1936–

DS771.R38   1987   951'.03   87–12910
ISBN 0-87332-441-2 (pbk.)

Printed in the United States of America

DAVID D. BUCK

Editor's Introduction

Although the Boxer Incident of 1900 remains a topic of considerable general interest in the West, the Boxers are not given a prominent place in Western interpretations of modern Chinese history. For example, the authoritative *Cambridge History of China* in volumes 10 (1978) and 11 (1980), dealing with the period from 1800 to 1911, does not contain a chapter devoted to the Boxer Movement; rather, the Boxers are covered as a less important phenomena in two chapters. Immanuel C. Y. Hsu emphasizes the movement's impact on China's foreign relations in a chapter dealing with late Qing diplomatic history, while Marianne Bastid-Bruguiere discusses the social character of the Boxers in her chapter entitled "Currents of Social Change." The Boxers are seen by most Western historians of modern China as a fascinating case of peasant antiforeignism that quickly faded, but whose activities had a considerable impact on China's foreign relations in 1900 and afterward.

The Boxers' role in modern Chinese history

In the history of modern China as written and taught in the People's Republic of China, the Boxer Movement has loomed much larger. Because of the approach to Chinese history that the Chinese Communist Party came to adopt in the late 1930s, peasant uprisings in general, and the Boxer Movement in particular, occupy a central place in the treatment of China since the mid-nineteenth century. In a short textbook, *The Chinese Revolution and the Chinese Communist Party* written at Yan'an in the late 1930s, Mao Zedong set forth the importance of the peasantry in Chinese history: "The scale of peasant uprisings and peasant wars in Chinese history has no parallel anywhere else. The class struggles of the peasants, the peasant uprisings and peasant wars consituted the real motive force of historical development in Chinese feudal society."[1] Thus, all instances of peasant uprisings, of which the Boxer Movement clearly is one, became subjects of great interest to historians.

The standard interpretation of modern Chinese history in the People's Republic of China has the modern era starting in 1839 with the First Opium War, but divides it into two subperiods, the Old Democratic Revolution (1839-1919) and the New Democratic Revolution (1919-1949). The latter begins with the May 4th Incident in Beijing when students protested China's unfair treatment in the Treaty of Versailles. In the New Democratic Period, the Chinese Communist Party becomes available to provide to the Chinese people, including the peasantry, a sound guide and appropriate strategy for revolution.

During the Old Democratic Revolution there are said to have been three "high tides" of revolutionary action: the Taiping Rebellion (1843-1865), the Reform Movement and Boxer Crisis (1898-1901), and the Republican Revolution (1905-1911). According to Marxist principles, this period should have produced a bourgeois capitalist revolution, and in fact finally did so in 1911, but the Chinese bourgeoisie, according to the Communists, always remained a weak and compromised class whose revolutionary will was not equal to their historical role. Evidence of these failings of the Chinese bourgeoisie and their representative intellectuals or politicians are given in terms of their continued subordination to the twin forces of domestic feudalism and foreign imperialism that had created and sought to maintain China's special status as a semifeudal and semicolonial country. A textbook on the Boxers written in the mid-1970s describes the leadership of the Reform Movement of 1898 in these terms: "The bourgeois reformists, who had just moved over from the landlord class, were dissatisfied with imperialist aggression and shocked and frightened by the national crisis [after 1895]. Nonetheless, their attitude differed from that of the people. They did not dare rise in revolution. Rather, they preferred to yield to the pressure of the imperialists in hope of getting support from among them for some reformist way of getting around the crisis."[2]

The Chinese bourgeoisie's failings, however, were offset by the staunchly revolutionary and increasingly nationalistic activities of the Chinese peasantry. Building on the tradition of peasant uprisings, the Taiping Movement came close to overturning the Qing dynasty in the middle of the nineteenth century, but the next important peasant movement, the Boxers, attacked the foreigners.

This shift of revolutionary action against the foreigners has been interpreted as basically a positive element in the Boxer Movement, for in the changing character of Chinese political and economic life in the last nineteenth century, foreign imperialism became the major force blocking progressive change in China. As Mao Zedong put it, "It is certainly not the purpose of the imperialist powers invading China to transform feudal China into capitalist China. On the contrary, their purpose is to transform China into their own semicolony or colony."[3] Thus the domestic feudal forces--the Qing aristocracy and the Chinese official and gentry elite who ruled in cooperation with the Manchus--became dependent on foreign imperialists to survive. The Boxers, who themselves never articulated much in the way of political analysis, have been interpreted as having understood the increased role of imperialist forces and to have struck out at the foreigners.

Certainly in the 1930s, the Chinese Communists put imperialism, meaning Japan in particular, in the place of the primary force blocking the proper historical development of China. The Communists saw the Chinese bourgeoisie as still subordinate to foreign interests and drew on the peasantry heavily in their struggle to stop Japanese aggression and to overturn the Nationalist government. The Boxers had much to recommend themselves as a model to China in the late 1930s. Their rabid antiforeignism contrasted with the less fervent stance of the Reform Movement leadership in the years from 1898 to 1901. The comparison of the staunch anti-Japanese nationalism the Communists sought to encourage in the North China peasantry in the late 1930s with the accommodationist attitudes of many urban residents was an obvious parallel to the Boxers and the Reformists.

The parallel between the Boxers and the positions adopted by the People's Republic of China after 1949 remained strong. Like the Boxers, the new government wanted to end the influence of Christianity, foreign capitalism, and those Chinese connected with foreigners. The new government in Beijing found itself confronting a different international force, in the form of the United Nations troops in Korea, but like the Boxers, the Chinese troops in Korea could not match the foreign troops' equipment and needed to call upon a superior will to gain victory. In 1958-59 with the Great Leap Forward and the subsequent break with the Soviet Union, the history of the Boxer Movement again could be cited as an example of the peasants' strong commitment to revolution as well as an example of Russian perifidy, though in a tsarist rather than a socialist guise. Finally, China, in a period of self-reliance, looked with approval on the Boxers' distaste for things foreign.

The Boxers had acted out of a sense of great injustice to themselves, combined with a feeling of impending disaster following China's defeat in the Sino-Japanese War of 1894-95 and the subsequent scramble for concessions among the Powers. They identified foreigners as the chief threat to China and without any qualms attacked these foreigners and those Chinese associated with them. The textbook from the 1970s gives special leadership to the peasantry in the struggle against imperialism in the years after 1895: "China's working class had not yet mounted the political stage. These masses of the people, with peasants as the main body, organized themselves to resist and fight crime-laden imperialism. It was they who felt most deeply, in their everyday life, the heavy weight of imperialism."[4] Thus the Boxers represented the larger interests of the Chinese people in their struggle against imperialism. These two characteristics--peasant leadership of a mass movement and the Boxers' antiforeignism--more than any others have made the Boxer Movement so prominent in the treatment of modern Chinese history in the People's Republic of China.

Characteristics of articles in this collection

Most of the papers in this collection were originally presented at an international conference on the Boxer Movement held in Jinan, Shandong, in November 1980 to mark the eightieth anniversary of the event. Many of those papers were revised and published in 1982.[5]

In selecting the articles for inclusion in this collection, I have attempted to provide a range of interpretative viewpoints, topics, and writing styles that reflect the general character of the historical profession in the People's Republic of China. Although these articles are marked by a tone of debate and differences, the reader will note that none of them really challenges the accepted nature of the Boxers themselves as traditional peasant rebels. The highly positive character of the Boxers' antiforeignism is now rejected in the era of Deng Xiaoping's "four modernizations" with the emphasis on extensive economic, political, and intellectual interaction with the rest of the world. Although the authors find some negative elements in the Boxer Movement, all of them reaffirm the Boxers' basic character, as indicated in the words of Lu Yao: "The Boxers were primarily an anti-imperialist struggle of the people. When we consider the complete nature of the movement . . . the Boxers retain their character of a traditional popular uprising and thus represent a continuation of that traditional form."[6] This means that the Boxers have both the strengths and the weakness of traditional peasant uprisings.

The views offered in this collection that really challenge the accepted characterization of the Boxers are those of Wang Zhizhong, who is so roundly attacked in the final selection by Sun Zuomin. What Sun finds so objectionable in Wang Zhizhong's articles are first the suggestion that the Boxers may not have acted on their own, but rather in response to the suggestions of the feudalistic officials and Court; and second his assertion that Boxer antiforeignism may have done more harm than good. Sun Zoumin is quick to see the implied challenge in Wang Zhizhong's interpretation to the whole notion of the peasantry being the great motive force in Chinese history. Sun's spirited defense, however, did not settle the issue.

In fact, the suggestions raised by Wang Zhizhong in his articles can be seen as the first salvo in what has developed into a major debate over the nature of history and historical writing in the People's Republic of China. The challenge to the orthodox views about the peasantry's importance have come from the Institute of Modern History in the Chinese Academy of Social Sciences, although it would be wrong to suggest that there is a unanimity of outlook among the historians there. Still, the leading historical journal in China, *Lishi yanjiu*

(Historical Studies), under the editorship of Li Shu and now Xu Zhongmian, has served as a major outlet for the views of the historians who have emphasized the examples of creativity and contributions from the Chinese bourgeoisie, but also have sometimes attacked the supposedly positive role of the peasantry.[7]

Following Wang Zhizhong's dissent from the prevailing views about the Boxers, the focus of attention shifted to the Westernization Movement (*yangwu yundong*). That movement began in the 1870s and saw Chinese entrepreneurs start up many new factories and use new means of production, usually under official sponsorship. Articles by Li Shiyue, Zhang Kaiyuan, Su Shuangbi, and others began to reflect a more positive and less condemnatory appraisal of those who led the Westernization Movement and to suggest that the movement had some positive results.[8]

The question became much more openly and heatedly disputed in 1984 with the publication of Li Shu's article in *Lishi yanjiu* entitled "Are the Masses the Creators of History?"[9] This piece has stirred up a debate that is still continuing in the pages of historical journals and newspapers. Li Shu's challenge is an intriguing one, for nothing in its approach departs from the basic principles of a Marxist historical outlook, but he argues that the scope of historians' concern must be more diverse than what led to the establishment of the People's Republic of China in 1949. He thinks the old approach is focused on this issue exclusively and misses much of what is important in terms of social change and development in China. Also, Li Shu argues that it was not always the masses who accomplished changes in history, for many important changes are accomplished by individuals rather than groups. Thus the individual and individual accomplishments warrant attention from historians. The ferment around this debate has been so great that it is referred to in the popular press as the "crisis in history" (*lishi weiji*).[10]

The eventual outcome of the debate and its effect on the research and writing of history in China remain unclear. However, peasant rebellions are still a valid subject of historical inquiry, and the peasantry and the "masses" continue to have their champions as the primary motive force in progressive change in China down to 1919.[11] The larger point, however, is that the beginnings of real differences in historical outlooks that appear in these articles--often in the form of apparent rather than real differences of interpretation--have given rise to a more fundamental debate in interpreting China's modern history in the general atmosphere of increasing intellectual liberalization and wider economic and social experimentation that has arisen in China during the 1980s.

Critique of the individual articles

Ding Mingnan's essay has been selected as the first in this collection because he provides a good general introduction to the prevailing questions asked about the Boxers and the standard evaluations of the movement. Ding is a fellow at the Institute of Modern History at the Chinese Academy of Social Sciences and a specialist in Sino-foreign relations and the Boxer Movement. He was elected the first chairman of the Society for Research on the Boxer Movement, formed in 1982, so he is much too modest when he states that he has not done much research on the subject.

Ding takes up four questions--origins, relations with the Court, the meaning of Boxer slogans, and the effects of the movement--that form the main issues debated most frequently in mainland historians' interpretations of the Boxers. The question of origins arises because of considerable differences that prevail. The idea, backed by some Western scholars, that the Boxers were developed from the militia is rejected out of hand by Ding and other contributors to this collection, but more serious debate arises over the question of secret society or heterodox religious influence on the Boxers. Ding states the dominant view among historians in the People's Republic that finds a strong secret society connection and distances the Boxers from the White Lotus religion.

This current interpretation departs considerably from views expressed by Fan Wenlan, Hu Sheng, and Qian Bocan in the 1950s, who emphasized the White Lotus influence on the Boxers. This secret society character, it should be noted, fits best with the conception of the Boxers as a traditional Chinese peasant movement, whereas the White Lotus origin and the militia origin theories do not support so well the idea of the Boxers as a patriotic, anti-imperialist movement led by the Chinese masses.

Ding's discussion of the Boxers' relations with the Qing Court moves carefully over this, the most difficult aspect of the Boxer Movement for those following the Maoist view of history. What makes this issue difficult is that the record shows clearly that Shandong officials as early as 1898 and the Empress Dowager in the spring of 1900 gave some support to the Boxers. Also, it is impossible to imagine tht the movement would have developed as quickly and become established in both Beijing and Tianjin in the spring of 1900 without some support from the top levels of the Qing state. The views advanced by Ding here, and supported with a fresh interpretation of the situation at the Court as advanced by Lin Huaguo in his article, portray the Qing Court and the Empress Dowager Cixi as essentially weak elements who clung to power through a series of compromises with the foreigners, the rabidly promonarchist elements of the Manchu aristocracy, and the Boxers.

Ding's discussion of the Boxers' slogans reveals how difficult it is to derive a coherent philosophy from the few materials available.[12] Ding's final concern with the effects of the Boxers emphasizes the point that Hu Bin makes in his essay about how the Boxers worked to prevent the break-up of China, in spite of their defeat by the eight-nation Allied Expeditionary Force. In closing, Ding affirms the close parallel that is seen to exist between the Boxer resistance to the foreigners in 1900 and the Communist-led resistance to Japan in the 1930s and 1940s.

Lu Yao's piece, "The Origins of the Boxers," is the longest and most detailed essay and shows his mastery of the materials on the Boxers. Lu Yao, too, does not stray from the standard interpretation, but he does an excellent job of sorting out the different strands in the Boxer movement and gives an account of their origins that emphasizes socioeconomic conditions and the secret society interpretation rather than the White Lotus interpretation. Lu Yao is an associate professor in the history department at Shandong University and has been involved with the work of gathering, editing, and publishing new materials on the Boxer Movement from field surveys, local records, and foreign sources. His articles usually have a tremendous amount of research behind each statement. His accounts are densely argued, and his articles repay careful reading.

Jin Chongji's essay addresses the issue of the connections between the White Lotus and the Boxers, a question that Lu Yao also described in detail. Jin Chongji, who is an historian employed at Cultural Relics Press (Wenwu chubanshe), concludes that the Boxers and White Lotus were distinct, but had some connections. He starts from the open, public nature of the village-based martial arts groups that gave rise to the Boxers and argues that these associations were fundamentally different from the White Lotus. Jin, however, backs away from his own evidence when he adds that the circumstances of early Boxer groups varied too much to permit generalization.

Although Jin Chongji presents what may seem like an elaborate account of the early stages of the Boxer Movement, it is worth noting that a much more detailed narrative of the early Boxer incidents is possible. For example, the dispute between the Harmony Boxers (Yihequan) led by Zhao Sanduo at Liyuantun (literally "Pear Tree Village") and the local Catholic church there can be described in much greater detail.[13] Jin does not inform his readers that the case stretches back to 1887 when a group of Catholic converts tried to take over the local Jade Emperor Temple (Yuhuangmiao) for their church. That temple had been built in 1861 and housed a school before falling into disrepair. The temple was dedicated to a Daoist deity whose veneration was widespread in the North China plain.

The chief site of the Jade Emperor cult was Mount Tai in central Shandong, one of China's sacred peaks. Various shrines associated with the Jade Emperor, but more particularly with the Goddess of the Azure Mists (Bixia yuanjun), a popular Daoist goddess since the tenth century, are located at the mountain top. In late Qing times she was commonly known as "The Old Woman of Mount Tai" (Taishan laonainai), who was said to be able to intercede to grant the birth of a son.[14] These temples were sites of great popular pilgrimages in late traditional times; even today thousands visit the mountain, now accessible by cable car, to seek the female deity's assistance in determining the gender of the officially sanctioned one child per family.

The dispute over the Jade Emperor Temple at Liyuantun went through a series of official investigations and dispositions, all intended to permit the Catholics to continue to occupy at least part of the former temple grounds while placating the outraged non-Catholic sentiment. None of these settlements was accepted, and the leaders of local martial arts societies, first Yan Shuqin and then Zhao Sanduo from a nearby village, marshalled demonstrations against the Catholics. In April 1897, ten years after the case began, three thousand people assembled for a three-day demonstration against the Catholic plans finally to build a church on the former temple grounds. Naturally an official investigation took place in 1897, and in a fracas between official forces and those of Zhao Sanduo one of the Boxer leaders was captured, returned to the district seat, tried, and executed. All of this occurred prior to what Jin Chongji describes as the "Liyuantun incident"!

Both Lu Yao and Jin Chongji want to draw a clear distinction between the Boxers and the White Lotus religious sects. The interpretation of this issue is the principal question on the early Boxer Movement. On the one hand, some conclusions must be reached about the links to the earlier White Lotus-connected uprisings of the eighteenth century and that of the Eight Trigrams (*bagua*) in 1813. This question has been explored in a somewhat different fashion by Susan Naquin in her book *Millenarian Rebellion in China* (1976) and in later articles, especially "The Transmission of White Lotus Sectarianism in Late Imperial China" (1985).[15] Professor Naquin describes the background of a particular leader of the 1813 rebellion, Feng Keshan, and shows that he was primarily a martial arts adept who joined others with a more distinct White Lotus sectarian background in activities that led to the uprising.[16]

In her article Naquin draws a distinction between two types of White Lotus followers: the strongly congregational tradition whose sutra-recitation sects lived quiet lives marked by regular services and abstemiousness, and the meditational tradition in which the believers were not tied to the congregational observances but practiced various forms of meditation as their chief religious activity. The evidence she cites about the meditational tradition, including emphasis on short, orally transmitted chants or mantras, lack of texts, absence of halls, and emphasis on proper performance of a few special acts in making religious observances (e.g., kneeling, bowing, kowtowing in a certain direction), all are elements that link the meditation tradition to the Boxers. Naquin leaves many questions unanswered, most importantly what was the ordinary relationship between the two traditions and why did the meditational tradition lead into the Boxers.

Although those questions remain unanswered, the advantage of Naquin's approach over those employed by Lu Yao and Jin Chongji is that she develops a general characterization of lines of activity within the White Lotus sects and gives them each several distinguishing marks, whereas Chinese scholars tend to make their distinctions based on the appearance of a single element in the Boxers as compared with the White Lotus. For example, Jin Chongji correctly notes that the Boxers operated public martial arts societies openly and without any apparent limits on membership for young men, but that alone is not enough to show that the Boxers did not derive from the White Lotus.

Further support for Naquin's idea of two varieties of White Lotus practice is found in the writings of Li Shiyu and in some of the field materials on the Boxers. In the latter some Boxer leaders refer to the differences between a civil (*wen*) and a martial (*wu*) tradition within the White Lotus and how they have striven to keep these connected, but still separate.[17]

Li Jikui's article, "How to View the Boxers' Superstitions," addresses the role of popular religion and folk beliefs in the movement. Li Jikui is a research scholar at the Sun Yat-sen Institute at Zhongshan University in Guangzhou. His approach is drawn from Marx and Engel's view of the peasantry, for he sees the peasants as having "intrinsic weaknesses that cannot be overcome by their own efforts." From this premise, he concludes the Boxers had to fail, and he attributes the causes of their failure to a combination of their religious superstition and the Qing Court's ability to mislead them.

Among the Boxers' positive characteristics, Li Jikui emphasizes how the spirit of nationalism had already penetrated into
the North China peasantry by the turn of the century and how
this nationalism served as a major goad to peasant action
against the foreigners. Li argues that these nationalistic ideas
were expressed through older patterns deriving from popular
religious practices. The most interesting elements of Li's article
are the long quotations from the folk tradition that has grown
up around the Boxers. Li uses those quotations to argue his
point about the lineage between nationalism and popular religion.

Li Jukui is not interested in the question of the role of religion in rural revolt generally. That broader question, which
could be discussed in terms of either Chinese or comparative
world history, has attracted great interest outside China. For
example, Janos M. Bak and Gerhard Benecke, in their book
Religion and Rural Revolt (1984), suggest that "great historical
religions contain two elements which can lead to their inclusion
into the sets of ideas that justify or mobilize revolt, namely
their eschatological promise and their mandatory system of
moral values."[18] The Boxers drew on both, for their opposition
to Christianity and foreign imperialism included a promise of
equality and justice for the Chinese peasant, plus a firm belief
that the practices of the Chinese Christians, the missionaries,
and the foreign governments who stood behind them all vio-
lated the basic moral and cultural principles of Chinese society.
As Li Jikui's article reveals, however, Chinese historians show
little inclination to compare Chinese peasant movements with
one another and remain silent on the wider question of com-
parisons with religious based rural revolt in other parts of the
world.

Qi Qizhang, a historian at the Shandong Academy of Social
Sciences, presents certainly the most complex line of argument
among these nine essays. He argues that the basic characteristics
of the Boxers changed markedly as the movement passed
through different stages in its history. Qi's scheme has a
twenty-month initial stage (October 1898 to June 10, 1900),
subdivided into three periods, followed by a brief "high-tide"
in the summer of 1900, and leading to a longer stage of decline

from mid-August 1900 until the end of 1902. He says that historians have often gone wrong about the Boxers because they have attributed characteristics of one stage to the whole movement. In fact, what Qi argues has a great deal of common sense behind it, for the character of the Boxers does seem markedly different at various stages in their history. Qi's categories, especially if one avoids the overclassification found in his subdivisions of the first period, meaningfully separate the stages from one another and describe the differences.

Only one of these essays deals with diplomatic history, but Hu Bin's certainly is the most highly polished piece of scholarship in the collection. Professor Hu teaches in the history department at Shandong Normal University in Jinan and has written a number of works on the late Qing, especially on the period from 1870 to 1900. His work is marked by strong nationalistic condemnation of foreign imperialism's effects on China as well as by denunciation of those Chinese, such as Li Hongzhang and Liu Kunyi, who cooperated with the foreigners. In some recent works, published since 1980, and especially in pieces written with Li Shiyue, Hu Bin has taken a more generous attitude toward the accomplishments of the Westernization movement.[19] His conclusion confirms the general consensus among both Chinese and foreign historians that the Boxers' fierce resistance gave an additional boost to the Anglo-American policy of cooperative imperialism toward China after 1900 and was a major factor halting the impending break-up of China.

Liao Yizhong's article, "Special Features of the Boxer Movement," along with that of Li Jikui discussed above, reflects most clearly the principles of Marxist historical thinking. Liao is a researcher at the Institute of History in the Tianjin Academy of Social Sciences and a prolific writer on the Boxer Movement. He is listed as chief author of the book *Yihetuan yundong shi* (A History of the Boxer Movement, [1981]), a recent standard treatment.[20] His essay translated here, "Special Features of the Boxer Movement," stresses the just nature of the Boxer struggle but points out the Boxers' own ideological weakness and leadership failings. As a result, Liao believes the Boxers were a much weaker movement than they have usually been understood to have been. What Liao is indicating here is a shift to a more balanced view of the Boxers than that which prevailed in the Cultural Revolution era (1966-1976), when the example of the Boxers' strong revolutionary will was often cited as a model for the present. In the 1980s, Liao dampens this high praise, without actually questioning the value of what Wang Zhizhong sees as the Boxers' blind xenophobia. Liao's article contains a number of references critical to the Boxers' xenophobia that reflect the general displeasure Chinese intellectuals feel about the antiforeignism of the Cultural Revolution. Yet, elsewhere in this article, Liao shows that he, like all of the others represented in this collection, basically accepts the standard interpretation of modern Chinese history that stresses the progressive and positive nature of peasant movements and contrasts it with the weakness and vacillation of the Chinese bourgeoisie.

**Translation of the terms "Yihequan"
and "Yihetuan"**

Considerable difference in understanding and interpretation result from the usage and the translations of the terms "Yihequan" and "Yihetuan." The name first used by the Boxers themselves was Yihequan or, as H. D. Porter, an American Congregational missionary who was residing at Pangjiazhuang in northwestern Shandong, translated it, "Righteous and Harmonious Fists." Here I have adopted a slightly different translation, "Harmony Boxers," based on the evidence provided by Lu Yao concerning the various boxing and militia organizations that existed around the Shibacun exclave in the late nineteenth century.

Porter's rendering is a literal and full translation of the name "Yihequan," which Zhao Sanduo and his supporters adopted sometime in the spring of 1898. Zhao stopped calling his group Plum Flower Boxers (Meihuaquan) when other groups using that name objected to the anti-foreign and anti-Christian actions led by Zhao Sanduo which they feared would lead to a general suppression of all associated martial arts groups. When Zhao Sanduo took the name Yihequan for his group he drew on the tradition of the 1770s in which there were White Lotus martial groups who used the term "Yihe." This tradition of using names suggesting harmony or concord remained strong in the region, as can be seen by the names adopted by the officially sponsored militia in the 1860s. These names, such as Yihe, Renhe, Peiyi, and Xieyi, all imply a just, common purpose to maintain public order and protect peasant lives and property. Lu Yao also details how these names were revived in 1896 for use by the militia. His account convinced me that a shorter translation of "Yihe" was warranted, and one that stressed the goal of public order; therefore I chose "Harmony Boxers."[21] Throughout the translations, "Yihequan" is translated as "Harmony Boxers."

The *quan* element in this name can be translated as "fists" or "Boxers." It refers to the practice of unarmed martial exercises that was the main activity of these groups. Drilling with weapons, even for the purposes of self-defense, was seen as improper by the Qing authorities, so such groups often selected names indicating they were unarmed. In fact, the Chinese peasantry could seldom afford real arms and had to make do with a few swords, knives, staves, or agricultural tools converted to military uses. The boxing associations engaged in some practice and drill with the available weapons, in addition to their greater emphasis on barehanded combat. The martial techniques of these adepts are the direct antecedents of Tai-chi ch'uan or other forms of Chinese martial exercise so popular in the West since the 1970s.

Sometimes the Chinese leave the term "Yihetuan" or "Yi Ho Tuan" untranslated, but I have followed standard practice and translated it here as "Boxers." In instances where some confusion might arise, I have included the original Chinese term in parentheses.

In English the term "Boxers" implies no sense of an official or semi-official group. In Chinese, however, the element *tuan* definitely does carry that suggestion. In North China generally, including Zhili's Wei district and the Shibacun exclave of Guan district, the Qing state had called upon local gentry after 1860 when it was carrying out the suppression of the Taipings and the Nian rebels to organize local militia for defense of their region. Even at the time not all the groups calling themselves *tuan* or militia were officially sanctioned. Later, when the militia were disbanded after 1870, some units might have continued an unofficial existence and still used their old designation as *tuan*. Lu Yao argues that the Boxers themselves began to use this appellation. His argument makes sense, and it suggests how powerful were such small indications of orthodox status in those days.

Marxist historians argue about the class character of local militia-style groups. They believe, with good cause, that local self-protection groups such as "united village associations" (*lianzhuanghui*) or unofficial militia (*xiangtuan*) were often dominated by wealthy and powerful local gentry whose economic and class interests were opposed to those of the peasantry. An important element in establishing the nature of any Chinese peasant movement is to determine its class character. The Boxers, especially those in Shibacun exclave and around Pingyuan and Chiping districts to the east, show little taint of landlord or gentry influence. The lack of a clearly articulated political plan, the large role of secret society practices, the coloration of heterodox religious elements, and the social background of most Boxer leaders all establish the Boxers as a true peasant movement.

The whole direction of interpretation of the Boxers in the People's Republic of China, however, has deemphasized official status for the Boxers and rather has stressed that the Boxers acted on their own. Consequently the Boxers character as a militia is constantly downplayed and explained in the most limited terms in order to preserve the movement's character as a popular peasant uprising. In Chinese this concern, in part, simply grows out of the name "Yihetuan," which implies militia connections; in English this can be avoided by simply translating "Yihetuan" as the "Boxers" and letting whatever evidence there is about official militia connections for the Boxers stand on its own.

The militia theory of the Boxers' origins

Lu Yao in his article refers to the fondness among foreign

scholars for the militia theory of origins. This argument runs that a group of officials with long service in North China, including Li Bingheng, Zhang Rumei, and Yuxian, successive governors of Shandong, encouraged the popular xenophobia within the province through the support of militia groups with an antiforeign character after the Qing dynasty's defeat in the Sino-Japanese War and in face of a growing threat of the dismemberment of the empire by foreign imperialism.

The case was put most strongly by George Nye Steiger in *China and the Occident: The Origin and Development of the Boxer Movement* (1927) in which he states that the White Lotus theory of the Boxers' origins as given by Lao Naixuan "must be rejected." He goes on to argue that "The so-called Boxers were a Tuan, or volunteer militia; they were recruited, in response to the express commands of the Throne, in precisely those provinces whose loyalty was most trusted" (p. 146). Steiger depends heavily upon the evidence of missionaries, both Catholic and Protestant, but especially on the account of Arthur Smith, an American Congregationalist missionary who had lived in North China for nearly thirty years at the time of the Boxer Movement. Smith and his boyhood friend, H. D. Porter, and their families maintained a mission station in the village of Pangjiazhuang, near the Grand Canal, in En district of Shandong province, near both the Shibacun exclave and the Pingyuan-Chiping centers of early Boxer activity.

Smith wrote that Boxers derived from "united village associations" (*lianzhuanghui*) who were active especially in the late winter to protect against robbers or other marauding bands. These villagers were armed with whatever weapons were available and displayed flags indicating their sanction by the local magistrate. He noted, "it is easy to perceive here the rudiments of a local militia," and then continued, "Village societies such as have been described, are constantly referred to as T'uan or Volunteers. Sometimes a day is fixed for the appointment of leaders for such in every township of the county, on which occasion the Magistrate entertains the local gentry and headmen at a feast. Early in the formation of the Boxer companies they began to style themselves, and to be styled by the Magistrates, "I Ho T'uan" [Yihetuan] or "Public Harmony Volunteers" which implied official sanction" (p. 174). Note, however, that Smith, who was well acquainted with practices in rural area, is unclear on the critical point of whether the designation *tuan* was self-appointed or official bestowed.

Steiger, however, adds evidence from Court edicts of 1896, 1898, and 1900 calling for the resurrection of the local militia in North China. Although the Chinese articles translated here do not pay much attention to this line of argument, Lu Yao does record that such a reconstitution of an official militia did occur in neighboring Wei district at the time the Boxers in the Shibacun exclave became increasingly active, and that some officials advocated enrolling the Boxers into militia units. Steiger's argument runs that the antiforeign officials used the imperial decrees for establishing militia as a screen behind which they could protect groups already charged with antiforeign acts or to encourage the growth of antiforeign groups among the peasantry.

A more mendacious interpretation of official reports by Yuxian and Zhang Rumei could be made in which they were using the Court's sanctions for militia as a cover for their own inability to control the unrest in the areas of their jurisdiction. They used the militia as a means of capturing control of unruly peasant groups and saving themselves from unwanted Court scrutiny. Lin Huaguo advances such an interpretation of the actions of Zhao Shuqiao and Gangyi at Zhuozhou in early June 1900, but he and others are reluctant to push such interpretations back into 1899 or 1898.

My own reading is that the Boxers were fundamentally a popular antiforeign uprising, but the government policy toward them contained some indications by certain antiforeign officials in Shandong--including Li Bingheng, Zhang Rumei, and Yuxian--that antiforeignism might be acceptable. These officials certainly lived in great fear of any major popular uprising because it was a frequent cause of official dismissal and disgrace. The Qing dynasty and its officials ordinarily suppressed all signs of peasant uprisings and, without Court sanction, it is difficult to believe that provincial-level officials in Jinan or their subordinates would ever actually encourage peasant outbreaks. Still, as good officials, they wanted to have matters in their territory firmly under their control, and granting militia status to unruly armed groups was a risky but not unknown means of establishing control. Also in the period after the Sino-Japanese War and the failed Reform Movement, the Qing Court was weak and its attitude unclear. Under these circumstances some officials acted in unusual ways.

Without challenging the overall conclusions of the Chinese scholars represented here, it does seem worthwhile to point out that there is evidence in the historical record that in addition to the dominant secret society character of the Boxers, their organization could use, possibly with official connivance, the militia tradition as a means of public organization for their antiforeign activities. The most extreme version of the militia theory of origins, which would have the Empress Dowager and a group of provincial officials stirring up the Boxers out of a peasant cultural distrust and then consciously unleashing these hordes against the defenseless foreigners, is obviously wrong. At the same time, it seems likely that some officials may have used the Court's militia decrees as a means to control the Boxers prior to 1900.

The purpose of this collection of translations, however, is to let Chinese historians speak for themselves about the Boxers, whom they have come to see as a major movement in the course of modern Chinese history. The articles offered here are only a small sample of the dozens of volumes and hundreds of essays published in the People's Republic about the Boxers, but it is hoped that the choices will help readers to understand the basic interpretations and outlines of the debates among historians about the Boxer Movement.[22]

Acknowledgments

This project has been too long in its completion, and I have to thank the translators' patience in the delayed appearance of their work. A similar expression of thanks is due to Douglas Merwin of M. E. Sharpe for his support.

I also wish to acknowledge the encouragement and assistance received from Professor Lu Yao in the genesis and completion of this project. My friend and colleague Garry Tiedemann has been a great source of information on the Boxer Movement, and his maps provided the basis for the work of Donna Schenstrom of the Cartographic Services Laboratory in the Department of Geography at the University of Wisconsin-Milwaukee. Li Guangli, who teaches at the School of Veterinarian Medicine in Changchun, Jilin, prepared the glossary. As for the translators, James Bollback teaches at Nyack College in New York; K. C. Chen is an associate professor of sociology at National Taiwan University's School of Law; Jay Sailey lives in Washington, D.C., where he works as an escort-translator for the United States Information Agency; and Zhang Xinwei is a professor of English at Shenzhen University in Guangdong.

I have adopted a common terminology for all the translations in dealing with many of the obscure and unusual terms surrounding the Boxer Movement. The responsibility for any inaccuracies on that or other scores in the translations resides with me as editor.

Notes

1. Selected Works of Mao Tse-tung (Peking: Foreign Languages Press, 1965), vol. 2, p. 308.
2. The Yi Ho Tuan Movement of 1900 (Peking: Foreign Languages Press, 1976), p. 11.
3. Selected Works of Mao Tse-tung, vol. 2, p. 310.
4. The Yi Ho Tuan Movement of 1900 (1976), p. 12.
5. Yihetuan yundongshi taolun wenji (Collected Articles on the History of the Boxer Movement) (Jinan: Qilu shushe, 1982). For a short report see Xu Xudian, "The 1980 Conference on the History of the Boxer Movement," Modern China 7, 3 (July 1981): 379-84.
6. Lu Yao, "The Origins of the Boxer Movement," p. 1 (typescript).
7. "Historians Seek New Perspectives," China Daily, September 23, 1986, p. 4; "Do the Masses Make History," China Daily, November 19, 1986, p. 4.
8. This debate is summarized in Jiang Jin, "Lishi yanjiu de feixianxinghua ji qi fangfalun wenti" (The Nonlinearization of Historical Research and Related Methodological Questions), Lishi yanjiu 1 (1986).
9. See Li Shu, "Renmin qunzhong shi lishi de chuangzao?" (Are the Masses the Creators of History?) Lishi yanjiu 5 (1984). For some of the ensuing controversy see Guo Ruixiang's reply and Li Shu's rejoinder in Lishi yanjiu 3 (1986); Zhu Masin's essay in Guangming ribao (Bright Daily), September 10, 1986, p. 5.
10. "Historians Debate 'Crisis' in Theoretical Methods," China Daily, May 12, 1986, p. 4; also see Xu Zhongmian's editor's note in Lishi yanjiu 1 (1986): 1.
11. See Muzi "Guanyu Zhongguo jindaishi jiben xiansuo wenti de taolun" (Discussion of the Basic Themes in Chinese Modern History) Renmin ribao (People's Daily), overseas edition, March 29, 1985, p. 3.

12. Ch'en Kuang-chung, one of the translators in this collection, completed a dissertation in sociology at the University of Illinois (1985) entitled "A Semiotic Phenomenology of the Boxers' Movement: A Contribution to a Hermaneutics of Historical Interpretation" (DA86000146), in which he attempts to use a new means of literary interpretation to unlock the meaning of the Boxers' slogans. For the most recent efforts in China see Chen Zhenjiang and Cheng Xuan, Yihetuan wenxian jizhu yu yanjiu (Annotations and Studies of the Boxers' Texts) (Tianjin: Tianjin renmin chubanshe, 1985). Chen and Cheng employ more traditional methodology than Dr. Ch'en's semiotics.

13. Lu Jingji, Yihetuan cai Shandong (The Boxers in Shandong) (Jinan: Qilu shushe, 1980); and Lu Yao, "Guan xian Liyuantun jiaoan yu Yihequan yundong" (The Missionary Case at Liyuantun in Guan District and the Boxer Movement), Lishi yanjiu 5 (1986).

14. For a report on the flourishing cult at Mount Tai from the late 1880s see Paul D. Bergen "A Visit to Taishan," China Recorder 19, 12 (1888): 541-46; also see description of the Azure Mists Temple in Zhongxiu Taian xianzhi (Revised Taian Gazetteer) (1927), juan 2, 67b.

15. Millenarian Rebellion in China: The Eight Trigrams Uprising of 1813 (New Haven: Yale University Press, 1976); Naquin's 1985 article is found in David Johnson et. al., ed., Popular Culture in Late Imperial China (Berkeley: University of California Press, 1985), pp. 255-91; also see her article "Connections Between Rebellions: Sect Family Networks in North China in Qing China," Modern China 8 (1982): 337-60.

16. Millenarian Rebellion in China, pp. 86-87, 90-92.

17. See Li Shiyu, Xiandai Huabei mimi congjiao (Contemporary Secret Religious Sects in North China) (Chengdu, 1948; reprinted Taibei, 1975); and "Yihetuan yuanliu shitan" (My Views on the Origins of the Boxers), Lishi jiaoxue (Teaching History) 2 (1979). Also Jin Chongji's article (typescript).

18. Manchester: Manchester University Press, 1984, p. 2.

19. See Hu Bin's excellent review essay in Zhongguo lishixue nianjian, 1984 (The Yearbook of Chinese History, 1983) (Beijing: Renmin chubanshe, 1984), pp. 105-13.

20. Liao Yizhong et. al. (Beijing: Renmin chubanshe, 1981).

21. See Don F. Draeger and Robert W. Smith Asian Fighting Arts (Tokyo: Kodansha, 1969).

22. For a detailed bibliography see Lu Yao, ed., Yihetuan yundong (The Boxer Movement) (Chengdu: Pashu shushe, 1985), pp. 518-66. This list contains all the materials published in the People's Republic of China from 1949 to 1983. For references after 1983 see the annual numbers of Zhongguo lishexue nianjian (The Yearbook of Chinese History), published by the Chinese Historical Association. Each annual number contains an essay on Boxer studies.

DING MINGNAN

Some Questions Concerning the Appraisal of the Boxer Movement*

The Boxer movement was a significant event in modern Chinese history that shook both China and the world. In the last several decades historians have done a great deal of research on the Boxer movement and have made some progress in explaining it. This is commendable and shows that people attach considerable importance to the Boxers. At the present time among historians different views continue to exist on nearly all of the important questions concerning the Boxer movement, including its origins, its characteristics, the meaning of its slogans, and its role in history. The disagreements over these points are considerable, but when the Boxer movement is approached as an academic question, and not a political issue, it is natural that such differences would exist. Only when we engage in free discussion and lay the facts on the table, speak rationally, and adopt the method of "letting a hundred schools contend" can we arrive a generally satisfactory conclusion. Yet, we all understand that in dealing with a legitimate academic issue it is permissible to continue our research and discussions, even if a general consensus is not achieved.

My own research about the Boxers has been negligible, so in this article I will only evaluate a few issues concerning the Boxers and present my own conclusions.

The origin and development of the Boxers

The debate over this question has gone on for a long time, but there are several main interpretations: first, the unofficial

---

*From Yihetuan yundong shi taolun wenji (Collected Articles on the History of the Boxer Movement) (Jinan: Qilu shushe, 1982), pp. 6-19. Translated by James A. Bollback with David D. Buck.

militia (*xiangtuan*) or official militia (*tuanlian*) theory; and second, the secret society theory, which includes both the possibilities that the Boxers were a secret society (*mimi jieshe*) or a secret religious sect (*mimi zongjiao, jiaomen*). I do not consider that the Boxers (*yihequan*) were a religious sect, and certainly they were not an official militia, but rather I see them as a typical secret society.

Originally, the Boxers' (*yihetuan*) predecessor, the "Yihequan" [often translated "righteous and Harmonious Fists," and translated here as Harmony Boxers] were related to the Eight Trigrams sect of the White Lotus system. The distinguishing characteristics of the Harmony Boxers were the practice of boxing, the incantation of magic charms, and possession by spirits. The earliest appearance of the name "Yihequan" was in 1778 (Qianlong 43) in the *Qing gaozong shilu* (Veritable Records of the Qing Dynasty for the Qianlong Reign [1736-1796]), where it is referred to as the "Harmony Heterodox Sect" (*yihe xiejiao*), meaning that it was considered a religious sect. In fact, it was not then a religious sect because it did not have a chief deity such as the White Lotus, who claim the Maitreya Buddha as their chief deity. The Harmony Boxers believed in many gods, more than we can name, including historical figures and folk heroes drawn from martial arts novels such as *Fengshenbang* (Canonization of the Gods), *Sanguo yanyi* (Romance of the Three Kingdoms), and *Xi you ji* (Journey to the West). Also, the Harmony Boxers did not have any sacred texts. Religious sects generally have secret texts, such as those of the Clear Tea Sect (*Qingcha menjiao*) entitled *Sanjiao yingjie zongguan daoshu* (A General Interpretation of Response to Kalpic Change According to the Three Religions). When the Qing government uncovered the various Harmony Boxer organizations, it was announced that there were some "evil books," but these were not sacred texts. In addition the Harmony Boxers did not have eight-character proverbs such as "True Home; Unborn Parents" (*zhenkong jiaxiang wusheng fumu*), which was used by the Tianli Sect.

In 1899 Lao Naixuan, the county magistrate in Wujiao, Zhili province, wrote the pamphlet *Yihequan jiaomen yuanliukao* (On the Origins of the Boxer Sectarians) in which he said that the

Yihequan was a branch of the White Lotus and was both a heterodox sect and a religious sect. Yet apparently Lao's characterization is untrue, for the Qing government always had strictly eliminated heterodox sects and punished the participants as rebels. Lao Naixuan cited the memorial of Zhili Governor-General Nayancheng written in 1808 (Jiaqing 13) to prove the point that the Harmony Boxers were a heterodox religious sect because, in addition to opposing Christianity, they were also antidynastic. His interpretation was intended to convince the government to suppress the Harmony Boxers as a heterodox sect. Zhili Governor-General Yulu and Shandong Governor Yuan Shikai both copied this pamphlet and disseminated it broadly, causing everyone to believe that the Harmony Boxers were a religious sect. That classification then became the basis for their suppression. Lao Naixuan's interpretation that the Boxers were a religious sect later was accepted by many people, but his interpretation will not bear close scrutiny.

The earliest occurrence of interpretation that the Boxers were an unofficial or an official militia was in a memorial written in June 1898 (Guangxu 24/5) by Zhang Rumei when he was governor in Shandong. He was also the first to use the term "Yihetuan." He reported to the Qing government that Harmony Boxers was an alternate name for the Plum Flower Boxers (*meihuaquan*), and that the people living in the Zhili-Shandong border region had a long-standing practice of boxing. These groups had been established "during the Xianfeng [1851-1862] and Tongzhi [1862-1875] reigns before there were Christian churches" and thus were popular organizations. The purpose of the Harmony Boxers was "to protect body and family and defend against bandits and thieves," and they "were not intended to find fault with Christianity." He also suggested that the government "turn this independent organization into an official one, changing the Boxers into a local militia (*mintuan*), thus taking the Boxers (*quanmin*) and enrolling them in the militia (*xiangtuan*)."[1] This memorial contradicted his own report of a month earlier in which he had said that in the Zhili-Shandong border regions "There are newly established societies called the "Righteous People" (*yimin*) who want to cause trouble for the Christians." That memorial included instructions to the local officials to take precautions against and strictly suppress these groups.[2] After Zhang Rumei had discussed these matters with Provincial Judge Yuxian, they had submitted the second memorial in June 1898 that emphasized the Harmony Boxers dated from the mid-nineteenth century. Thus the Boxers could not be an anti-Christian organization because they were established before any churches had been organized in Shandong. During the 1850s and 1860s every province was commanded to establish local militia because the dynasty needed them to fight against the anti-Qing forces of the Taiping and Nian. Zhang's second memorial sought to show that the Boxers were derived from those militia organizations. It is clear that Zhang Rumei's memorial was intended to protect the Harmony Boxers and to avoid further investigation by the Qing government. In his book *China and the Occident*, the American George Steiger said that the Harmony Boxers were "neither a revolutionary nor an heretical organization; it was a lawful and loyal volunteer militia,"[3] but that view does not have a reliable foundation.

In a book published in Taiwan by Dai Xuanzhi, *Yihetuan yanjiu* (Researches on the Boxers), he states that the Boxers were an official militia that came "from the Plum Flower Boxers who in turn originated from the Harmony Boxers." He adds, "The transformation into the Boxers (*yihetuan*) began in 1897 (Guangxu 13) at the time of the missionary case at Liyuantun in Guan county, Shandong"[4] [Translator's note: in the Shibacun exclave]. Dai Xuanzhi's principle evidence is Zhang Rumei's memorial, but because this memorial itself contains some doubtful material, Dai's statements about the local militia naturally are even less believable.

A proper evaluation of the Boxer movement must take into account its origins and development. A local militia was a force supporting the decaying control of the Qing dynasty, while historically secret societies had been actively suppressed by the Qing dynasty. Secret societies and local militia were in opposition, and I have concluded that the Boxers (*yihetuan*) were a secret society and not a militia.

Research about the Boxers' origins and development is difficult. First, the records from the Qianlong [1736-1796] and

Jiaqing [1796-1820] periods are scattered. In the seventy-odd years between 1820 and the Sino-Japanese War of 1894-95, there is almost a complete absence of records about the Boxers. Second, people outside an organization find it difficult to understand its internal affairs, so if one is not a member of a secret society it is very difficult to understand it. Twenty years ago comrades in Shandong and other places carried out some local investigations, visiting and interviewing elderly people who had participated in the Harmony Boxers. Some of their materials are quite valuable. Recently some comrades have used the records of those investigations and have made some progress on the questions of the Boxers' origin and development. If we want to clarify further this difficult problem, still deeper investigations are required.

The relationship of the Qing government to the Boxers

This question is related directly to one's evaluation of the Boxer movement. There are great differences of opinion about this question. Harold Vinacke wrote of the Boxers, "From their first appearance they had strong support at the Court, as also among the officials in the provinces, but it was not until the siege of the legations had commenced that the Court finally threw its official support on the side of Boxerism."[5] At the time the Boxer Protocol of 1901 was signed, the representatives of each country wanted the Qing government to admit its mistake in supporting the Boxers and also wanted it to apologize while paying indemnities to the imperialist nations. But, we may ask, did the Qing government really support the Boxers' anti-imperialist struggle? The answer is certainly no.

We know that from the mid-nineteenth century onward, the Qing government had signed a series of unequal treaties with the various Powers in which they accepted the responsibility of providing protection for the foreigners, including the obligation of protecting Christianity. On behalf of the Powers, the dynasty became a manager of the Chinese people, who were reduced to slaves of imperialism. The masses throughout China joined the anti-Christian struggle, often attacking the foreign forces that encroached upon China. The Qing government was reprimanded often by the imperialists for those attacks, and these reprimands had an influence on the relationships between the Chinese and foreign reactionaries. For a long time the Qing government always strictly suppressed the anti-Christian masses and did as much as possible to satisfy the demands of the foreign invaders. For example, in the 1890s after the Chengdu and Juye missionary cases, the Qing Court went so far as to dismiss Sichuan Governor-General Liu Bingzhang and Shandong Governor Li Bingheng. Prior to the Boxer movement, the Qing government increased greatly it slavishness toward imperialism; consequently it was impossible for the Qing to support the masses in their anti-Christian struggle. Therefore it is unreasonable to say that the Qing strongly supported the Boxers.

In suppressing the Boxers, Qing government orders formed the basis of the activities by the governors of Zhili and Shandong. *Zhidong jiaofei diancun* (Telegrams Concerning the Suppression of the Boxers in Zhili and Shandong), a compilation made an official in the Zhili governor-general's office, provides evidence of this.[6] This collection preserves telegrams concerning the suppression of the Boxers exchanged by Yulu, Zhang Rumei, and Yuxian and shows that there were differences in attitude about carrying out the Qing Court's orders among these officials. Zhang Rumei and Yuxian believed that in the conflict between the Christians and the people of Shandong, the Christians should be held completely responsible, but they were local officials and thus could not disobey the Court's orders, so the best they could do was to claim that the Boxers were a local militia in order to shield it from the Court. Yao Luoqi, a leader of the Boxers in Guan county, was executed by Zhang Rumei. Zhu Hongdeng and the monk Benming, Boxer leaders in the area of Pingyuan and Chiping, were captured and executed by Yuxian. These are all established facts.

Until June 15, 1900, Governor-General Yulu suppressed the Boxers. Just two days before the Dagu forts were occupied he had memorialized the Qing government to "give the order to send high officials to suppress and punish the Boxer bandits" (*quanfei*).[7] After the battle at Dagu began, he believed it was impossible to suppress the Boxers and to resist the foreign columns at the same time, so rather than fighting on two fronts he requested permission from the Qing government to let him "use the Boxers."[8] Because Yulu had favored suppression of the Boxers from the beginning, the British government, even after the fighting commenced, sent instructions to the British forces in China stating that if Yulu requested asylum from the British he should receive it.

Because Yuan Shikai advocated suppression of the Boxers, some imperial censors denounced him and called for his dismissal from office. In those circumstances he did not dare to carry out openly a large-scale massacre of the Boxers, but, based on the orders from Beijing to "suppress and disband the Boxers first" and "to capture the leaders and disband the followers," he captured and executed the important leaders Wang Liyan, Yu Wenyi, and others. Consequently the Boxers' anti-imperialist struggle became stronger in Zhili but weaker in Shandong. The belief that the Qing government strongly supported the Boxers cannot be proven. Similarly, it cannot be proven that officials from other provinces supported the Boxers.

When the Empress Dowager became the leader of the Qing government after the failure of the Hundred Days Reform of 1898, the dynasty's ruling power was greatly weakened. A conflict emerged with the imperialists over the succession, and Cixi suspected that the imperialists intended to use their military might to support the Guangxu emperor. At this critical juncture, just when China was about to be sliced like a melon, the Qing dynasty was unable to gain the support of the people in order to suppress the Boxers. Some imperial censors repeatedly memorialized the Empress Dowager with the argument that it was extremely dangerous to suppress the people and support Christianity. They pointed out that although the Boxers were causing trouble, they were after all Chinese, whereas the loyalty of the Christian converts was owed to their churches and they had become "foreigners" (*yangmin*).[9]

The Qing dynasty's policies toward the Boxer vacillated between strict suppression and more moderate disbandment. Consequently it was unable to carry out any policy effectively. In the circumstances, the Boxers flourished. In addition, by raising the banner "Support the Qing, Eliminate the Foreigners" (*fu Qing mie yang*), the rapid development of the Boxers was aided. Yet, in spite of the conflict between the Qing dynasty and imperialism, the dynasty could not risk depending on the Boxers and declaring war against the imperialists. Not long after the fighting began, the Qing government cabled its representatives abroad to inform them of the situation and described the situation in China accurately. The message stated: "China herself is not strong and cannot fight against all the foreign nations at the same time. Nevertheless we cannot rely on a rebellious mob (*luanmin*) to conduct a war against the Powers."[10] Yet, in mid-June the dynasty went ahead and declared war against the imperialists. The only possible explanation is that the Boxers forced the dynasty to do this, and that the overall development of the situation compelled the Qing to act in this fashion, for the outcome was contrary to the Qing government's basic intent.

China had suffered the encroachments and oppression of foreign states for many years. In 1900 when China faced an immediate crisis of being carved up by the imperialists, the Boxers' anti-imperialist struggle received the overwhelming support and sympathy of the masses. A Qing official wrote: "From the beginning, the Boxers' (*quanmin*) ideas received popular support in the form of money and grain. . . . Wherever they went people would contribute grain for their support and join up. No one could stop these actions, not even the advice of a father, an elder brother, a wife, or a son. Their minds were filled with thoughts of killing the enemy and obtaining victory."[11] This statement is short, but it clearly reflects how the masses wanted to join the movement.

The spread of Boxers took place extremely rapidly, precisely because the masses supported the Boxers' anti-imperialist struggle. In mid-April 1900, a small number of Boxers appeared in the community at Lugouqiao (Marco Polo Bridge), in the countryside near Beijing. By the end of May their numbers had swollen to an army of tens of thousands. They occupied Zhuozhou and tore up sections of the railway between Lukouqiao and Baoding. Boxer groups from all over gathered together, mainly in Beijing and Tianjin.

They became a strong force, threatening the control of the Qing dynasty. In early June the situation had already become quite serious. After the British consul met with the president of the Zongli Yamen, Yikuang [Prince Qing], on June 5, he reported back to the British government that Prince Qing admitted that soldiers were disobeying orders to the extent that the Qing government could no longer guarantee the safety of Beijing.[12] Grand Councillor and Grand Secretary Ronglu sent a secret telegram to Zhang Zhidong later in June in which he spoke of the Boxer's great strength. He wrote: "Even in the two imperial palaces and in the residences of the officials half of the people are members of the Boxers. Most of the regular soldiers, both Manchu and Chinese, belong to the Boxers. The Boxers in the capital swarm like a cloud of locusts and are impossible to control."[13] As the residents and part of the military troops stationed at Beijing became members of the Boxers, the regular government actually lost control of the situation. If it had continued to insist on the suppression of the Boxers, it would have been destroyed.

After June 10, Admiral Seymour's expedition was proceeding along the Tianjin-Beijing railway in an attempt to invade Beijing, while the mass movement opposing the imperialists' military aggression became more highly developed. Robert Hart, British head of the Chinese Imperial Maritime Customs Service, early on had pointed out that "The Court appears to be in a dilemma: if the Boxers are not suppressed, the Legations threaten to take action--if the attempt to suppress them is made, this intensely patriotic organization will be converted into an antidynastic one."[14] The American capitalist scholar H. B. Morse, wrote: "If she would not be submerged by the Boxer wave, she must ride it--that, if it was not to destroy the throne, it must be turned against the foreigner."[15] Since the Qing government did not want to be destroyed, it was forced to allow the Boxers to enter the city of Beijing. After having considered the matter several times, the Qing recognized the Boxers as legal and declared war against the imperialists. To protect itself, the dynasty adopted this policy as a temporary strategy. A little more than a week after declaring war, the Qing government directed its ambassadors to guarantee to the foreign nations that it would "devise a means to punish" the Boxers. It also continued unceasingly to appeal to Russia, Japan, Britain, France, Germany, and the United States to end the war. Ronglu was ordered to dispatch someone to the foreign legations for discussions on ceasing hostilities, and Li Hongzhang was named the peace negotiator.

All of this activity by the Qing was to no avail because the dynasty could no longer control the situation, while the imperialists had made up their minds to occupy Beijing. In the end, the Empress Dowager had to flee and the Allied forces occupied Beijing. As the Empress Dowager was escaping, she issued an edict calling for "the suppression of the bandits" and ordered the official troops to suppress the Boxers, requiring that all members be killed. This is how the Qing dynasty sold out the Boxers.

The overall situation during the development of the Boxers shows clearly that the Boxers were in opposition to the Qing dynasty and that the declaration of war against the foreign governments was forced upon the Qing. China's feudal rulers did not and could not support a patriotic movement of the people. The idea that the Qing Court and officials in some locations had supported the Boxers from the beginning is not verified by the facts.

The slogan "Support the Qing and Eliminate the Foreigners"

Evaluation of the slogan "Support the Qing, Eliminate the Foreigners" is important when appraising the Boxer movement. This slogan, which contained in a nutshell the outline of the Boxers' activities, was quite popular and widely used. Some comrades have argued mistakenly that this was merely a local usage; others are equally mistaken when they divide the slogan and emphasize "eliminating the foreigners" without mentioning "support of the Qing." The whole slogan reflects the strongly nationalistic character of the Boxer movement. Other places, such as Dazu in Sichuan, Changyang and Changle in Hubei, or Hainmen and Taiping in Zhejiang, had similar slogans at this time.

When we look closely at the slogan, we can see that it did not require the overthrow of the Qing dynasty, and also that it did not oppose the feudal land system. Thus we can reasonably conclude that the Boxers were not antifeudal. But, in those days when China occupied a semicolonial status, and the Qing dynasty had become the tool of the imperialists who controlled China, it was impossible to attack imperialism without also attacking feudal rule. Seen in this light the Boxer movement was anti-imperialist, but at the same time it also served to lead the attack against feudalism. We might say that subjectively the Boxers were not antifeudal, but objectively they did have some antifeudal effects.

The "Eliminate the Foreigners" part of the slogan basically is xenophobic. Antiforeignism is related to xenophobia, but they have different characteristics, even though their objectives are similar. We can admire the Boxers' anti-imperialist aspects, but xenophobia goes so far as to exclude anything that comes from a foreign county, or from foreigners, or new ideas from the outside. Xenophobia entails a great blindness. Proceeding from the lowest cultural level, xenophobia engenders a narrow outlook and is the sign of ignorance and backwardness. There are some historical reasons why circumstances produced this phenomena, but we should not make excuses for the Boxers, and even less should we advocate or support such xenophobia, because that would turn back the clock of history to advocate primitivism.

Still, the Boxers' xenophobia does requires some analysis. Some comrades incorrectly believe that isolated events, such as the burning of foreign merchandise or a change of a place name, became widespread, daily events. Others incorrectly argue that the Boxers' attacks on trains, railway lines, telegraph lines, and machinery were examples of Ludditism or xenophobia. The Boxers did many of those things, but there is no evidence that they destroyed machinery. The destruction of sections of the Beijing-Baoding railway was done to stop the transfer of Nie Shicheng's troops from Tianjin to Baoding, while similar actions along the Tianjin-Beijing railway not only halted the advance of Nie Shicheng's forces, but were even more important in blocking the advance of Admiral Seymour's

Allied Forces. The Boxers destroyed the telegraph lines so that Seymour would be cut off from Tianjin, and indeed his troops were forced to retreat in the end.

There was yet another reason for the Boxers' destruction of the railroad. Engels has written: "British capitalist power wanted to construct railroads in China but China's railways will mean the destruction of the entire basis of the small landholding peasant economy and the household industry; because there is no large industry in China to create a balance, one million people will have no means of earning a living."[16] It had not been long since China first began to build railroads, but events nevertheless verified Engels's prophetic words. Yuan Yong, a high-ranking official who was killed during the Boxer movement, said that the Boxers destroyed the railroad and the rolling stock because "the laborers from countryside around Beijing had lost their livelihoods from carts, boats, and shops because of the railroads. As a consequence more than forty thousand people put on red clothes and yellow bands [to join the Boxers]."[17] From this we can see that the Boxers' destruction of the railroads had military as well as economic grounds.

Some comrades, while they agree the Boxers were an anti-imperialist movement, do not accept that the Boxers really intended to "support the Qing" as their slogan suggests; I disagree. There is no question that anti-imperialism is patriotic, but we cannot say that the Qing dynasty, which had already sold out China, was truly patriotic. Lenin has emphasized that the concept of a nation is a notion about which people hold a variety of contradictory ideas. We would be wrong to think that the people of that time could differentiate between the Qing dynasty and the Chinese nation, particularly at a moment when they were aroused and confused. Therefore, I cannot accept the explanation that the slogan "support the Qing" meant only wholehearted support and protection for the Qing dynasty, for it also contained the meaning of supporting and protecting China. Seen in this fashion, there is not much to this question of the Boxers' support for the Qing.

For the Boxer movement, this slogan "Support the Qing, Eliminate the Foreigners" was quite effective. Initially it mobilized and drew large numbers of the masses to join in the anti-imperialist struggle. Some landlords, feudal intellectuals, and even some officials joined the Boxers because of it, thus indicating its considerable appeal. On the other hand, this slogan caused the masses to relax their vigilance against the Qing dynasty's schemes and thus became a means by which the Qing sold out the nation. In that respect, it was harmful to the Boxer movement.

There were other slogans during the Boxer movement in addition to the widely used "Support the Qing, Eliminate the Foreigners." For example, in the town of Linzhen in Fucheng county of Zhili, the Boxers used the slogan "Proclaim the Spirits, Assist the Sect, Eliminate the Foreigners, Cooperate with the Boxers" (*chuanshen, zhujiao, mieyang, gonghe yihetuan*).[18] This is quite unusual and worth our notice. By our ordinary understanding of these words, this slogan would put them in opposition to established authority and indicate they intended to overthrow the dynasty. Unfortunately our documentary records on the Fucheng Boxers are deficient, and we have no means at present of further investigating this unusual slogan.

The effects of the Boxer movement

Another important question in evaluating the Boxers is the issue of what historical effects the movement had. Some say the Boxers' attacks on imperialism prevented the Powers from occupying and dividing China; others argue it was the contradictions among the imperialists that stopped the partition of China, and not the Boxers. These two interpretations have basically different emphasis. My conclusion is that the struggle by the Boxers was the primary reason that imperialism could not partition China. The contradictions among the imperialists were also a factor, but these were only secondary.

Before the Boxer movement, China already faced a serious threat of partition. The Powers occupied the ports, fought for special railway and mining rights, and marked out spheres of interest. All of these can be considered preparations for partition. During the Boxer period the strong resistance by Chinese soldiers and people taught the wild imperialists that any division of China would bring with it serious consequences. The American Congregationalist missionary Arthur Smith said that the failure of Admiral Seymour's expedition to reach Beijing would dispose of "once and for all the favorable proposition so often advanced that it would be possible for a small but well-organized and thoroughly equipped foreign force to march through China from end to end without effective opposition."[19] The resistance of the Chinese soldiers and people during the movement destroyed the notion that "foreign armies were invincible" and let them know that whoever wanted to divide China would pay heavily. The commander-in-chief of the Allied Forces, Count von Waldersee, admitted that "no matter if it were the United States, European, or Japanese forces, none of them had sufficient intelligence or strength to control one-fourth of the world's population. . . . Therefore, the policy of partition was not feasible."[20] During the Allied occupation of Beijing, Robert Hart wrote a number of articles that gave the reasons why China could not be split up. He wrote: "As regards partition--that plan, like every other, has its good and bad sides; but, with such an enormous population, it could never be expected to be a final settlement, and unrest, unhappiness, and uncertainty would run through all succeeding generations." He added, "a partitioned China will make common cause against its several foreign rulers," and would everywhere "show the existence and strength of national feeling."[21]

The former U.S. consul in Beijing, Charles Denby, wrote: "Among the countries of the world, China is the one most unsuited for partition. There is no other race that is as similar, as united, as bound by ancient ties and enchantments as the Chinese. . . . Those European Powers who desiring to partition China would use force to split her up . . . would only be met by unceasing resistance. . . . In the end, the subjugators would fight among themselves, producing a great battle that would bring disaster to the one that loses."[22]

From these quotations we can reach two conclusions. First, China's partition would have moved the Chinese people to an unceasing opposition; second, partition would have created conflict among the imperialists.

At the time the Boxers were defeated and the Qing Court

had fled, the imperialists thought that with their military strength the partition of China might be inevitable. The reason they did not dare to partition China was not that they feared the Boxers might rise again to do battle, but rather because the Boxers' struggle showed that the Chinese people had an un-quenchable fighting spirit which they must respect. Because they lacked both "the intellectual strength and military might" to rule China directly, the imperialists altered their plans to partition China. Thus, we could say that the Boxers prevented the partition of China by the imperialists.

The contradictions within the imperialist camp remained, but such contradictions among Britain, France, Germany, Belgium, Portugal, and Spain had not prevented the partition of Africa a few years earlier. There is no sound reason to believe that these contradictions were strong enough to prevent the partition of China. The mutual disagreements among the imperialists was only a secondary cause, while the imperialists' fear of the Chinese people's opposition became the chief factor preventing China's division. There were two causes for this outcome, but we should be clear which was the primary one.

Although the Boxers did not make antifeudal demands, they did attack the Qing government and thereby weakened the dynasty's control while also increasing the masses' consciousness. The Qing government lacked strength to suppress the Boxers, revealing its weakened condition. When those in the North de-clared war, the southern provincial officials sided with the imperialists in calling for "the mutual protection of the southeast region" thereby revealing the division in the government. At one point the Qing recognized the Boxers as good people (*yimin*), but shortly thereafter it sold them out, revealing how shameless they were. Just after the "peace proposals" were received, the Qing proclaimed the need to "estimate China's material strength and match it with our desires." This stance unequivocally demonstrates that the Qing had become the complete slave of imperialism. The Qing dynasty's complicity helped clear up the masses' misconception and to improve their understanding of the Qing. As a consequence the masses greatly strengthened their resolve and trust and finally determined that the Qing government could and should be toppled. A decade later this occurred.

The Boxers were a spontaneous anti-imperialistic movement of the masses at the turn of the century of which the peasants formed the main supporters. It had the specific characteristics associated with that period and that class. The Boxers burst upon the scene just as the crisis over partition had reached its peak, when China's dismemberment seemed at hand. The Boxer movement occurred only just in time to save the people facing the crisis. The Boxers were a product of the increasingly deep encroachment by the imperialists against China, and they constitute the high point of the struggle by the Chinese people against encroachment and partition. They also reflect the long anti-Christian struggle among the Chinese people.

The Boxers were a spontaneous struggle of the peasantry. They did not have a unified organization, or a central leadership, or a unified program. At the time the proletariat had not yet appeared on the political scene; the capitalists not only refused to lead the Boxers, but even cursed and despised them. Initially the feudal ruling class supported the Boxers, but then it switched and began denouncing them. As a consequence, the Boxers, who had received feudal support initially when they fought fearlessly against imperialism, then found themselves fighting alone.

Although involved in a brave struggle, the small producers' own individualism and narrow-mindedness blinded them and produced serious negative consequences for their movement. Their defeat became inevitable. Some people say the Boxers had a good organization and sophisticated regulations, along with sound strategic thinking--even superior to that of the proletariat--but that does not tally with the facts.

Some interpreters emphasize the feudal nature and xenophobia of the Boxers, but that too is an overemphasis of certain qualities. Still one point must be stressed: the Boxers' anti-imperialism was much more advanced than that of the Taipings. Furthermore, the Boxers surpassed the bourgeoisie's reformist and revolutionary factions. The peasants had learned the evils of imperialism from the realities of life. They understood that imperialism was the enemy of the Chinese people. Although the Qing Court's reform faction referred to Russia as "the nation that destroyed people," the Qing dynastics still looked upon the British and the Japanese--who also were imperialists--as their friends based on the belief that the British and the Qing could become allies. The Chinese bourgeoisie, who traditionally had a childish understanding of imperialism, mistakenly thought they could use the imperialists' sympathy and support for the benefit of the Chinese revolution. This faith reached the point that the warlord government after 1911 in its "Foreign Affairs Manifesto" continued to accept the unequal treaties and continued to pay indemnities to the foreign nations, while it also protected foreign firms' profits. These methods only invited the imperialists' disdain and taunts. Although the Boxers were not equal in a struggle against imperialism, they still dared to fight. We should admire the Boxers for this because the bourgeoisie never thought about undertaking a face-to-face struggle with the imperialists. They lacked the determination required for such a contest; in fact they were in no position even to talk about such a struggle.

Naturally the Boxers had many weaknesses, failures, and mistakes. Their movement ended in defeat, but when we point out their shortcomings, we must apply scientific analysis and historical understanding to affirm their historical effectiveness. The Boxers' blood was not spilled in vain, for "their brave struggle became one of the foundations of the Chinese victory fifty years later." Still, today, this evaluation seems correct.

Notes

1. Guojia danganju Ming-Qing danganguan (Office of Ming-Qing Archives of the National Archives), eds., Yihetuan dangan shiliao (Archival Materials on the Boxers; hereafter DASL) (Beijing: Zhonghua shuju, 1959), vol. 1, pp. 14-15.
2. Ibid., p. 14.
3. China and the Occident: The Origin and Development of the Boxer Movement (New Haven: Yale, 1927), p. 146.
4. Yihetuan yanjiu (Taibei: Wenhai chubanshe, 1964).
5. H. M. Vinacke, A History of the Far East in Modern Times, 4th ed. (New York: Crofts, 1945), p. 162.
6. Lin Xuezhen, ed., "Zhidong jiaofe diancun" (Telegraphs Relating to the Suppression of the Boxers in Zhili and Shandong), in Yihetuan yundong shiliao congbian (Collected Materials on the Boxer Movement) (Beijing: Zhonghua shuju, 1964), part II, p. 70.

7. DASL, p. 143.

8. Ibid., p. 158

9. Ibid., pp. 42-43, 46, 53, 60.

10. Ibid., p. 203.

11. Ibid., p. 178.

12. Edmund Wehrle, Britain, China and the Anti-Missionary Riots, 1891-1900 (Minneapolis: University of Minnesota, 1966), p. 165.

13. Rong Wenzhonggongji (Collected Works of Ronglu), juan 4, p. 1.

14. Item 1711, letter of May 27, 1900, in J. K. Fairbank et al., eds., The I.G. in Peking: Letters of Robert Hart, Chinese Maritime Customs (Cambridge: Harvard, 1975), vol. 2, p. 1230.

15. The International Relations of the Chinese Empire (London: Longman's Green and Co., 1910-18), vol. 3, p. 247.

16. The Complete Works of Marx and Engels (Chinese ed.), vol. 38, p. 467.

17. "Luanzhong riji cangao" (Fragments from Diaries Written During the Days of Disorder), in Qian Bocan et al., eds., Yihetuan (The Boxers) (Beijing: Shenzhou guoguangshe, 1951), vol. 1, p. 347.

18. Cheng Zongyu, ed., Jiaoan zouyi huibian (Selected Memorials on Missionary Cases), juan 7.

19. China in Convulsion (New York: Fleming Revell, 1901), pp. 443-44.

20. Yihetuan, vol. 3, pp. 86, 244.

21. These from the Land of Sinim (London: Chapman & Hall, 1903), pp. 49-50, 97.

22. China and Her People (Boston: Page, 1902), vol. 2, pp. 92-93.

LU YAO

The Origins of the Boxers*

The problem of the Boxers' origins has constituted a major
concern for foreign scholars in their studies. In the last few
years, we Chinese scholars also have begun to take up this
question again. A review of the evidence can help us understand how the controversy arose over this matter, while also
helping us to plumb the character of the Boxer movement itself.

The Boxers were primarily an anti-imperialist struggle of
the people. When we consider the complete nature of the
movement--their class composition, the development of their
struggle, the sources of their organization, and their political
ideas, the Boxers retain their character of a traditional popular
uprising and thus represent a continuation of that traditional
form. This essay examines the affiliations and derivations of
the Boxers, but it also discusses how it retained the characteristics of a traditional peasant movement. Elsewhere I have
published an article, "Some Questions Concerning the Initial
Stages of the Boxer Movement" (1979), in which I presented my
ideas on the questions of origins.[1] They have not changed
greatly, and so I have corrected, supplemented, and refined
those earlier views but have not altered my basic conclusions.

Over the past thirty years Chinese historians' views about
the affiliations and derivations of the Boxers have been quite
diverse. Some say the Boxers derive from the Harmony Boxers
(Yihequan), while others see them as a combination of the Harmony Boxers and the Big Sword Society (Dadaohui). Another
view asserts the Boxers gained the support of the masses be-

---

*From Yihetuan yundong shi taolun wenji (Collected Articles on the History of
the Boxer Movement) (Jinan: Qilu shushe, 1982), pp. 65-97. Translated by K. C.
Chen with David D. Buck.

cause of their connections with the Spirit Boxers (Shenquan).[2] Still others think the Boxers were an amalgamation of the Big Swords, the Red Boxers (Hongquan), the Spirit Boxers, and the Harmony Boxers.[3] Finally there are those who believe the Boxers were a combination of the Big Swords, the various boxing associations, and the White Lotus sect.[4] In addition there are diverse views about whether or not the Big Swords and the Boxing groups should be classified as secret religious sects, as secret societies, or as legally constituted self-protection groups.

These differences about the organizational basis of the Boxers leads obviously to contradictory views about their origins. There are three basic positions. The first holds that the Boxers were a branch of the White Lotus sect (*bailianjiao*). The historians Fan Wenlan and Hu Sheng supported this view soon after 1949, yet there are differences in this school over which branch of the White Lotus developed into the Boxers. The most common view is that the Boxers derive from the Eight Trigrams sect (Baguajiao) within the White Lotus tradition. Others point to particular subsects of the Eight Trigrams, arguing variously for the Li Trigram,[5] a combination of the Qian and Kan trigrams, or simply the Kan trigram subsect.[6] Another position is that the Boxers represent a combination of the White Lotus sect with the Eight Trigrams.[7] A second, completely different line of reasoning holds that the Boxers did not necessarily derive from religious groups at all, but grew out of clandestine societies that developed around the practice of martial arts.[8] The third approach holds that the Boxers lacked a single origin. Instead, this view holds that some Boxers were connected with the White Lotus sects, some were linked with secret societies, and still others were members of popular organizations for self-protection based on martial arts exercises. Also, among foreigners there are many who argue that the Boxers originated from militia groups, but among Chinese scholars few take that position.

These contradictory viewpoints reflect the complex nature of the Boxers, while revealing that the available historical evidence is insufficient to settle these questions. This essay combines various sources--documentary evidence, archival evidence, and the results of oral history interviews--in order to explore further this question of origins. It is most useful to separate the question into two parts, affiliations and derivation, but often these two separate issues are considered to be one. The affiliation issue involves the various boxing associations of the late nineteenth century that participated in the Boxer movement, while the question of derivation leads back to the Qianlong reign (1736-1796) to investigate the boxing associations and religious sects in the eighteenth century. In this essay I take up the affiliation question first and then the derivation issue. I believe that either line of investigation leads to the conclusion that the Boxers were connected both to secret societies and to secret religious sects. Moreover, these characteristics prove that the Boxers belong to the category of traditional peasant uprisings.

The rise of the Boxer movement and its affiliation with secret societies and secret religious groups

The Boxer movement was a spontaneous but intermittent mass struggle that lacked unified organization and leadership. If we confine our attention to the early phase of the Boxer movement in Shandong, however, we see that it was not a riotous crowd completely lacking in organization, but rather developed from secret religious sects and secret societies.

There are three areas in Shandong where the Boxers began or that had an influence on the Boxers' beginnings: (1) Shan and Cao counties in Caozhou prefecture in the southwestern part of the province, (2) Chiping county in Dongchang prefecture, and (3) Guan county in Dongchang prefecture, both in the northwestern part of the province. Contemporary analysis of the Boxer Movement gave two different accounts that reflect these geographical differences. One traces the Boxers to incidents involving the so-called Eighteen Militia (Shibatuan) in Guan county; the other linked up these same disturbances in Guan county with those occurring to the south in Caozhou.[9]

The 1896 anti-Christian struggle in Shan and Cao counties was a direct forerunner of the Boxers. The Big Sword Society, which initiated this struggle, was an organization combining secret society and secret religious sectarian elements. Two years later, in 1898, the two centers of Boxer activity in Shandong both were farther to the north. Then the movement linked the activity in the Shibacun exclave of Shandong's Guan county with other Boxer groups operating to the east in Shandong's Changqing, Chiping, and Pingyuan counties. Elements from the Guan county exclave were led by Zhao Shanduo and Yan Shuqin, while Zhu Hongdeng led the Spirit Boxers in the second area. Zhao Sanduo led groups that originated from a secret society, while Zhu Hongdeng's groups combined secret society and religious sectarian origins.

The Big Swords of Cao and Shan counties were established by Liu Shiduan, who came from the village called Shaobing Liuzhuang and derived from the tradition of Cao Deli of Caolou village in Shan county. The society was first known as the Armor of Golden Bell (Jinzhongzhao), but in its public activities it was known as the Big Sword Society. Liu Shiduan's organization derives from the Golden Bell of the Jiaqing reign (1796-1820) and has no connection with the Straight Knife Society (Shundaohui) of the Qianlong and Jiaqing periods. His organization, originally named the Golden Bell by Liu Shiduan, had links with the Kan Trigram subsect of the Eight Trigrams and constituted an offshoot from the Kan Trigram.

There is considerable evidence linking Liu Shiduan's Big Swords to the Kan Trigram. First, Liu's group was set up initially with help from a White Lotus devotee named Zhao Jinhuan who hailed from Hejian prefecture in Zhili. This Zhao Jinhuan also was an adherent of the Kan Trigram. In Cao and Shan counties most members of the Big Sword were also members of the Kan Trigram groups, although some were members of Li Trigram groups. No links to other Trigram subsects have been found.[10]

Second, the distinguishing characteristics of the Cao and Shan Big Swords were the ingesting of charms and chanting of incantations, a belief that they could become invulnerable to swords and other blows, and the practice of breathing exercises to make them invulnerable to gunfire. The charms and incantations they used had a distinct White Lotus character. Moreover, Liu's Big Swords honored a Venerable Master (*zushiye*), a practice that derives from the White Lotus tradition. Third, when Liu Shiduan started his group he claimed its goal was protection against local banditry, but its true nature was rebellious and antidynastic.

According to recollections we gathered during fieldwork investigations,

> Liu Shiduan organized the Big Swords, also called the Golden Bell. That year there was a drought and the grain harvest was poor. The local people became excited and, fearing starvation, soon became angry. In southwestern Shandong they gathered under the banner of the Big Swords and assembled in a great meeting near the towns of Anling and Gudui [in neighboring northern Jiangsu]. Liu dressed himself in an opera costume and appeared riding a horse and carrying a sword, all the while proclaiming himself emperor. He wore yellow, the imperial color, and rode in a palanquin. After the wheat harvest it rained heavily and the mob returned home to plant a summer crop of soybeans. There was a folk rhyme that recalled those incidents that went,
> "A banner was raised on Anling hill,
> but the rain washed it away."
> All of this took place prior to the anti-Christian outbreak of 1896 in our area.[11]

We can connect those events with the earlier Eight Trigrams activities of 1813 in Cao county. In November 1813 (Jiaqing 18/10), the Shandong governor Tongxing sent troops to Dingtao to suppress the Eight Trigrams. Forewarned of this assault, the sect members "all fled into the neighborhood of Jujiaji in Cao county where there was a place called Anlingji where most of them were killed or captured."[12] Liu Shiduan's Big Swords started in his own native village, and later he led his followers to assemble at this same Anling hill in nearby northern Jiangsu where he raised a flag of revolt, donned an opera costume, and proclaimed himself emperor. All of these elements parallel the 1813 Eight Trigrams revolt.

In the situation that existed following the Sino-Japanese War of 1894-95, the crisis in China was more severe and the power of imperialistic Christianity had greatly increased, so the masses in the Cao and Shan area organized under the Big Swords to resist Christian aggression. In line with this, the Big Swords became more overtly anti-Christian. Later, following the execution of Liu Shiduan, Cao Deli, and their followers, the Golden Bell gradually cut their ties with the Kan Trigram and extended their activities into Shandong around Yanzhou, Yizhou, Caozhou, Jining, and Dongchang prefectures. At this same time in Caozhou prefecture another secret society, known as the Red Boxers (Hongquan) or Red Brick Society (Hong-zhuanhui), appeared, but apparently it had no connections with the Eight Trigram subsects, for it was simply a secret society and operated independently.

The Harmony Boxers (Yihequan) in the Guan and Wei county areas comprised two separate elements. Those led by Yan Shuqin practiced Red Boxing, while those who followed Zhao Sanduo were the Plum Flower Boxers (Meihuaquan). There is no evidence to suggest that prior to May 1899 either of them had any relations with religious sectarians. Zhao San-duo began public demonstrations of Plum Flower Boxing around Ganji, a market town in the Guan county exclave, in 1896, and thereby turned what had been a secret ritual into a public one. Still, that fell short of making the Plum Flower Boxers a legally acceptable militia body. In the eyes of the Qing government these boxing associations still were illegal groups, and their public activities were possible only because of the inability of the officials representing the feudal Chinese states to suppress them.

Sometime around 1898 Zhao Sanduo changed the name of the Plum Flower Boxers into the Harmony Boxers (Yihequan) [often translated "Righteous and Harmonious Fists"][13], but his organization was different from the former Plum Flower Boxers or the Harmony Boxers that had existed in Guan county before 1813, for his was martial in nature and had no religious coloration. Thus Lao Naixuan was wrong to link Zhao Sanduo's Harmony Boxers with the Li Trigram subsect. Lao Naixuan's conclusions were based on memorials written by the Nayan-cheng in 1815 that revealed a connection between the Harmony Boxers of 1813 and the Li Trigrams. Even as Lao Naixuan himself stated, this identification was merely a deduction without conclusive proof.[14] My previously published views, which indicated that Zhao Sanduo was a religious practioner and linked his Plum Flower Boxers with the Kan and Qian Trigrams, were partially mistaken, because in its initial stages his movement clearly was not a heterodox religion. The Harmony Boxers from Shibacun exclave of Guan County did not become subordinate to the Kan Trigram until 1900.[14]

Lao Naixuan's arguments to prove that the Harmony Boxers were heterodox, such as the claim that they practiced "divine possession," martial arts, and secret incantations to protect themselves against guns, and that they used the titles venerable master (*zushi*), elder (*dashixiong*), and junior elder (*ershixiong*),[16] in fact reflect characteristics only of the Spirit Boxers of Chiping and the Golden Bell of Caozhou prefecture, and were not found among Guan county's Harmony Boxers. Sectarian elements did not appear among Zhao Sanduo's followers until May 1900 (Guangxu 26/4) after a secret meeting at the Temple of the Great Buddha (Dafosi) at Zhengding in central Zhili and the subsequent banding together of the Harmony Boxers with the Iron Shirts (Tiebushan) and Golden Bell sectarians from Zhili's Dongguang prefecture.[17]

The Spirit Boxers (Shenquan) from around Chiping began publicly teaching their ideas early in 1895 but did not start any assaults until the spring of 1899.[18] The Spirit Boxers remain relatively unstudied. Documents indicate that these Spirit Boxers' earliest appearances were in Ying county of Zhejiang province in 1765 and in Shangqiu county of Henan province in 1777. They do not appear in other places. Then, between 1820 and 1860, the Spirit Boxers cropped up in Daxing county of Zhili as well as in their old home, Yin county in Zhejiang. Nothing concrete is recorded from the eighteenth century about the distinguishing characteristics of the Spirit Boxers, but the official court history for the period mentions the "Spirit Boxing Society" (Shenquanhui).[19] In the Xianfeng reign (1852-1862) it was recorded that the Spirit Boxers around Ying county believed in incantations.[20] The characteristics of divine possession, charms, and incantations appeared only in the Chiping Spirit Boxers after the Sino-Japanese war. In fact, it was only after Zhu Hongdeng arrived in Chiping and became the leader of the Spirit Boxers that this organization became linked with the Eight Trigrams.

There exists incontrovertible proof that Zhu Hongdeng had links with the Eight Trigrams. First, Jiang Kai, the magistrate of Pingyuan county, stated, "He [Zhu Hongdeng] styles himself as Heavenly Dragon (*tianlong*)."[21] The title "Heavenly Dragon" appeared in association with the Eight Trigrams sect around 1811 after Fang Rongsheng, head of the Return to the Origin Sect (Shouyuanjiao), compiled two texts using the term Heavenly Dragon in their titles, "Tianlong babu yuanming ce" (The Eightfold Understanding of the Heavenly Dragon) and "Longhua dingguo ce" (The Truth of the Dragon Flower).[22] In 1861 the White Lotus sects around Guan, Xin, and Tangyi counties used the name "Eight Trigrams Sect of the Heavenly Dragon" (*Tianlong bangua jiao*). Also in that same year the Qing dynasty troops at Zhangguanzhai in Guantao county "captured the agent Li Daxiao ('Big Flute'), who confessed to being the head of the illegal Eight Trigrams Sect of the Heavenly Trigrams."[23] Second, in Zhili province's Gucheng county, which borders on Dezhou in Shandong, there was a monk Dagui who also was a boxing master and confessed after his arrest to being a "Junior Master (*shidi*) to Zhu Hongdeng."[24] This Monk Dagui had been a member of the Li Trigram subsect.[25] Third, the battle reports from Gangzi Lizhuang on October 11, 1899 (Guangxu 25/9/7), that record the engagement between Zhu Hongdeng's Spirit Boxers and the Qing government troops stated that Zhu Hongdeng wore a red winter cap with ear flaps, red leggings, and marched at the head of his columns with two red flags. His followers' weapons were decorated with red cloth. All of their markings were red, to indicate the color of fire and also to distinguish them from other trigrams.[26] The basic meaning of Zhu Hongdeng's surname is red; it also was the name of the deposed native Chinese Ming dynasty (1368-1644). His given name, "Hongdeng," means Red Lantern. Our oral history investigations contain evidence that when Zhu Hongdeng arrived in Chiping he was wearing red cotton trousers and a red cotton lined jacket.[27] All of the above elements prove that he was a disciple of the Eight Trigrams sect.

Fourth, after Zhu Hongdeng came to Chiping there was a piece of doggerel that spread in the region:

This suffering is not so bad,
If to twenty-four, fifteen you add.
When the Red Lantern sign descends on the morrow
Then we'll all truly know sorrow.[28]

Contemporary documents record the same item with a slightly different line order, indicating that the rhyme was widely known at the time that Zhu Hongdeng operated in the area. The various versions indicate the rhyme probably was derived from planchette writing.[29]

In Chiping a rumor spread among the people that "After Zhu Hongdeng, then will come the Red Lanterns," and there are recorded instances in which Zhu Hongdeng is referred to by the name of the first Ming emperor.[30] Zhu Hongdeng's name as well as the name of the Buddhist monk Benming (one meaning of which would be "Source of the Ming dynasty") indicate that the leadership of the Spirit Boxers had the goal of "overthrowing the Qing and restoring the Ming."

Fifth, our survey around Chiping and Pingyuan revealed that the Spirit Boxers' regulations included injunctions such as "be filial and loyal; respect elders and care for the young, be amicable at home and in your village"; or "Don't steal; Don't commit adultery"; and "Honor your parents, respect your brothers and be amicable at home."[31] Each of these comes directly from the Codes of the Eight Trigrams in existence prior to 1813. For example, in 1786 the Eight Trigrams' follower He Jiande confessed, "I was a child when I joined the Eight Trigrams sect, which had been propagated by the Liu family for generations. They taught good behavior, including respect for Heaven and Earth, filial devotion to one's parents, being peaceable in your home area, and before eating to raise one's hands in praise to Heaven and Earth."[32] In addition there were four proscriptions, including "Don't watch improper behavior; Don't listen to improper stories."[33] Thus it is not speculation to say that Zhu Hongdeng was a member of the Eight Trigrams sect. In fact, it is obvious that the Spirit Boxers under his leadership had become linked with the Eight Trigrams.

Now the question remains what branch of the various boxing associations had the greatest influence on the Boxers when they appeared in Shandong. These various boxing associations often were unrelated and certainly lacked unity of any sort. So the question becomes, which among them can be called the dominant one? The answer is the Spirit Boxers, who had connections with the Golden Bell.

In these years at the end of the nineteenth century, when the Spirit Boxers first appeared in Chiping, their practices involved only divine possession and incantations prior to 1899 and did not include invulnerability to cuts or other injury, nor did they employ breathing exercises. Only after they had established connections with the Golden Bell tradition from the Caozhou region of Shandong in 1899 did those latter characteristics appear in Chiping. Thenceforth the various altars (*tan*) established in the Tianjin and Beijing region by the Boxers all were derived from Spirit Boxing groups whose practices show the influence of the Golden Bell. The proclamations entitled "Exhoration to the People to Remain Peaceful and Not Establish Heterodox Groups" or "A Proclamation of Four Means to Suppress the Boxers" issued by Shandong Governor Yuan Shikai especially directed their opprobrium to the Spirit Boxers and the Golden Bell. The boxing association leaders, the Monk Wuxiu and Wang Zingyi, who operated in the area around the counties of Zaoqiang, Jingzhou, and Gucheng in Zhili, taught Spirit Boxing, and it was primarily the Spirit Boxers who occupied Zhuozhou in Zhili during June 1900. Later, when the Boxers spread to Beijing, most of the boxing groups there were related to the Spirit Boxers or the Golden Bell. Sometimes these were even called "Golden Bell Militia" (Jinzhongzhaotuan).[34]

As we have seen, the Golden Bell began in Cao and Shan counties in southwestern Shandong, while the Spirit Boxers arose in Chiping and Pingyuan farther to the north in that same province. Later on the Boxers from those two areas were referred to as "old units" (*laotuan*). In August 1900 the Boxer leaders Zaixun and Gangyi sent troops to Cao district and to Dongchang prefecture in Shandong to recruit new members for these old units. It is obvious these "old units" were important in the Boxer movement. Once the Boxers held sway in Peking there were many Boxer units not connected to the Spirit Boxers or the Golden Bell, but rather controlled by the gentry or created by imperial decrees. Still, those new elements were small in number and limited in their influence. The most important Boxers at Peking were these "old units," especially those from Shandong. They were derived from the Golden Bell practices and maintained anachronistic forms of peasant struggle based on secret society and religious sectarian traditions.

The members of these early organizations primarily were poor peasants. In the two centers of the Boxers in Shandong we can demonstrate this. In Chiping the official archives record, "I [Pi Yuan] believe that the area under my jurisdiction around Chiping and Boxing is a refuge for these Boxers (Yihetuan). The weather in my region has been exceptionally dry and the numbers of the poor have increased. When these poor people assemble they all claim to be Boxers. The majority of these Boxers are poor people without any means of livelihood."[35] It was in Shibacun enclave at Liyuantun in Guan county that the Boxers first appeared. When the local magistrate attached the fields of the local boxing practitioners in the fall of 1900, he reported, "These Boxers are mostly homeless people without any means of livelihood. Yan Shuqin and 'Little Pock-Mark' Gao, both of whom have already been executed, did not have any property or other means; . . . the twelve households connected with Xi Desheng, who also has been executed, altogether owned 140 *mu* of land. All of it was ordered confiscated and sold at auction."[36]

The property of the so-called Eighteen Heroes (Shibakui) was even less. Only one was a landlord with more than 100 *mu* of land. The others were from families that held only a few *mu* or tenants who had to work for wages in order to make ends meet.[37] Most of the adherents and members of the Boxers were poor laboring people. The leaders of the Boxer struggle in Shandong were "poor peasants, servants, unemployed poor, or agricultural laborers. Others were boat haulers on the Yellow River, ferrymen, water carriers, teamsters, carpenters, umbrella repairers, kitchen workers, and street vendors in various lines

such as pancakes, stationery, bamboo, tobacco and miscellaneous goods, as well as dyers, blacksmiths, innkeepers, and out-
of-work bookkeepers, soldiers, and teachers."[38] In Zhili and
other provinces the Boxers adherents were much the same as in
Shandong.

Why did so many landless or nearly landless poor peasants
and handicraft workers join this movement? Everyone knows
that the Boxer movement was based on anti-imperialism and
was caused by the increasing imperialist aggression that
threatened the Chinese people. Yet from this single premise it
is difficult to understand the full nature and true character of
the Boxers. From the conditions in Shandong we can see that
the movement was a product of the serious contradictions
existing with China and a reflection of a deep social crisis as
well. Much has been written about the former, but the latter
warrants some additional comments.

If we return to the situation around Chiping we find there
were more boxing groups in that area than any place else in
Shandong. In 1899 there were "more than 800" boxing associations, while Chiping county itself was commonly referred
to as "a county of 860 villages"[39] Why were there so many
boxing groups around Chiping? The basis of these groups cannot be explained adequately by the aggression or oppression
from the Christians. Of course these were factors, but in the
summer of 1898 there had been a major flood in the area following a breach in the Yellow River dikes. Tyrannical administration added to the suffering caused by this natural disaster
and was an additional factor that caused the widespread activities of the Spirit Boxers. The *Chiping xianzhi* (Chiping
Gazetteer) states,

On August 11, 1898 (Guangxu 24/6/24), the Yellow River
breached the eastern dike south of Xiangshan, flooding
Chiping. The raging waters covered even the higher ground
with several feet of water, flooding all the roads and turning
the land to the east into a great marsh. Almost all of the
people's houses and goods were lost and they became
refugees. In this predicament they survived several weeks of
rainy weather, unable to cook, living in wet clothes and

going without meals for days at a time. These flood conditions lasted for a long time, until after the mid-autumn fes-
tival [September 30, 1898] before the water receded. The
suffering of the refugees was unprecedented.[40]

The important areas where the Spirit Boxers and the Harmony
Boxers were found, such as Changqing, Qihe, En, Pingyuan,
Xiajin, Guan, and Qiu counties, all had similar conditions.
These tragic circumstances were manmade for the most part.
Breaches of the Yellow River dikes had been occurring for
several years as a consequence of embezzlement of flood control funds by officials of all ranks. By 1898 the dikes had been
in misrepair for many years and there existed a pattern of annual flooding. The censors in their impeachment memorials had
reported this corrupt administration of the Yellow River Con-
servancy for many years. It was this pattern of government
tyranny combined with natural calamities that produced the
starving masses and the gangs of vagabonds who increasingly
joined the boxing associations in northwestern and southwestern
Shandong. Court records about Xiajin county for 1900 state
that "In spring and through the summer, the rainfall was small.
Those short of food became agitated and their numbers
increased steadily." In Guantao county it was reported, "This
year (1900) from spring through summer the rainfall was un-
seasonably small and the wheat harvest was meager while the
fall crops were not planted until too late. . . . The people
became disturbed and those short of food took action to divide
the grain stores." In Daming and Yuancheng counties in Zhili,
"Because of the failure of two wheat crops, the wealthy
households have been forced recently to divide up their grain
stores with the poor."[41]
    In Zhili's Wei county, "This year [1900] had been bad. The
poor have no means to survive and they flock to the boxing
groups (*quanmin*) which use the slogan 'equal division of the
grain' to justify their anti-Christian activities."[42] Around
Shunzhi that year, "The rain was slight and the wheat crop
never fully headed, so the people became unsettled. In Shanxi
province since last winter there has been an even more serious
drought that has made people fear for their very survival. It is

said this caused the starving people to become involved in disturbances."[43] The pages of *Wanguo gongbao* (Intelligencer of All Nations, founded by Timothy Richard) emphasized that there were many more destitute people than normal and pointed out that in North China, "Poor people lack livelihood and the weather has brought on a famine. The weak topple in the roadside ditches while the stronger become outlaws. The loss of their hope to live a normal life gave rise to the disorder and now they advocate dividing the wealth among rich and poor."[44]

Thus, we can see that the great majority of those who were part of the Boxer Movement were famine victims or homeless wanderers. This characteristic reveals that theirs was a traditional peasant movement. It was no coincidence that this struggle "to distribute the grain equally" appeared in many separate locations.

Connections between the Boxers and the White Lotus during the eighteenth and early nineteenth centuries

In the section above we have discussed the Boxers' affiliation, and in this section I want to examine their derivation. What connections did the Boxers have with the White Lotus active since the eighteenth century in this same region of North China? We can see some similarities and differences simply in the boxing groups' organization and in their distinguishing characteristics.

To argue that the Boxers carried no elements of the Eight Trigrams sects simply contradicts the historical facts; yet if one follows Lao Naixuan's theories to claim that the Boxers were the same as the White Lotus of the eighteenth century, that is even more a distortion of the historical truth. My conclusion is that the Boxer movement and these earlier White Lotus peasant uprisings both manifest the trait of joining together various boxing groups such as the Red Boxers, the Spirit Boxers, the Harmony Boxers, and the Golden Bell with the Eight Trigrams and its associated sects in order to advance their struggle. Starting with this premise, I believe the Boxer movement carried on the tradition of the earlier White Lotus peasant uprisings. Even more, I argue that the late nineteenth century boxing groups all evolved in areas that were pockets of earlier activity. So from this premise, I want to examine the "prehistory" of the Boxer movement.

We know that White Lotus uprisings of the Qianlong and Jiaqing periods (1736-1821) involved both secret societies and secret religious sects. Sometimes these two were distinct, and sometimes they had affiliations. Often they joined together into one movement. The Qing administration declared the Harmony Boxers, the Plum Flower Boxers, the Red Boxers, the Iron Shirts, the Golden Bell, the Straight Swords, the Tiger-tail White (Huweibian), and other boxing groups to be "heterodox associations" (*xiehui*) in the nature of secret societies. The various religious sects linked with the Boxing groups such as the Return to the Origin (Shouyuan), the Eight Trigrams, the Greater Vehicle (Dacheng), and the Clear Water (Qingshui) all were branches of the White Lotus. The Qing administration strictly proscribed these secret religious sects: "According to the codes, leaders of heterodox sects who worship concealed images and incite the people shall be sentenced to death by slow strangulation." Another part of the Qing code stated, "Those who follow heterodox sects that cause disturbances among the people shall be banished to Muslim cities as workers or slaves among the Muslims."[45] Boxing or martial arts associations that did not fall into the category of heterodox religious sects still were prohibited: "There is a proscription against establishing an association for the purpose of practicing martial arts or collecting weapons."[46] "Those who are self-proclaimed teachers of martial arts, practitioners, or students of these arts are all liable to a punishment of one hundred strokes and banishment to a distance of 3000 *li*."[47] "Idlers, the unemployed, self-proclaimed teachers of the martial arts, and their followers are liable to punishment of one hundred strokes or banishment for three years."[48]

We can see that simply practicing in the boxing groups was not permitted. Those who practiced martial arts were liable to considerable punishments. Consequently, being involved in the indoctrination of others into these boxing groups took on the character of joining a secret society.

How should we distinguish between a secret society and a secret religious sect? In 1900 when suppressing the Harmony Boxers, the Qing administration pointed out, "those [groups] deriving from earlier White Lotus sects all employ incantations to delude people, even today."[49] Thus we may conclude that these incantations are the distinguishing characteristic of heterodox religious sects (*xiejiao*) and mark the distinction between them and heterodox associations (*xiehui*).

For example, the White Lotus in their activities during the century prior to 1820 often took the form of a secret society rather than their true nature of a secret religious sect because the Qing state had promulgated comparatively lighter penalties for these secret societies. Records reveal that various boxing associations such as the Straight Swords (Shundaohui), the Red Boxers, the Spirit Boxers, and the Harmony Boxers in the past had linked up with heterodox religious sects such as the White Sun sect (Baiyangjiao) and the Clear Water sect to carry out anti-Qing activities. In 1757 it was said, "the White Lotus and the Lo sects are declining, but recently there is an association known as the Straight Swords that appears everywhere. This association has no scriptures and does not practice vegetarianism. All one needs to join is a sword."[50] In the year 1774 it was recorded, "Shandong has a long tradition of martial arts which include groups such as the Harmony Boxers and the Red Boxers."[51] In that same year the Clear Water sect uprising of Wang Lun occurred. Wang Lun was an adherent of both the Clear Water sect and the Harmony Boxers, showing that it was quite common for these two groups to join forces in an uprising. "These various religious sectarians, all of whom derive from the White Lotus, call themselves 'Harmony Boxers' and incite the ignorant country folk. They bring harm and are hateful."[52]

The location in which these Harmony Boxers of the eighteenth century had been most active was the very location, Guan county, where the Boxers (Yihetuan) began in the late nineteenth century. According to historical records, "The Qing administration discovered in 1778 that "in Guan county in Shandong there is a family surnamed Yang who have brought together many others and established a Harmony Boxing sect

(Yihequan xiejiao). They recruit villagers who after paying 300 to 500 cash are taught the techniques of boxing." These Yangs were the family of Yang Sihai from the village of Wanerzhuang in Guan county, and the teaching of Harmony Boxing dates from Yang Shihai's father.[53]

In the Jiaqing period (1796-1821), and particularly during the Tianli sect uprising of 1813, it was common to have boxing groups involved with the Eight Trigrams. Feng Keshan, leader of the Li Trigram subsect, was a disciple of the Red Boxers and practiced their martial arts. So was Zhang Jingwen, the leader of the Li Trigrams, who had been a member of the Red Boxers. The center of their activities was one of the locations where the late nineteenth century Boxer movement started, Caozhou prefecture in Shandong. "Around Caozhou we find the people are obstinate and participate in both the Harmony and the Red Boxers. These groups use force to intimidate the weak and are a great local scourge."[54] In addition, in 1797-98 (Jiaqing 2 and 3) at Ganji market town in Guan county a Golden Bell group appeared under the direction of Zhang Lojiao:"The secrets of the Golden Bell that Zhang Lojiao studied permitted him to be struck with a sword on the shoulder or on the side without injury. Such blows did not draw blood or leave any marks."[55] Thus we can see that already at the end of the eighteenth century, the Golden Bell had acquired the distinguishing characteristic of promising to render its adherents invulnerable to weapons. This distinguished the Golden Bell from other boxing groups. We have shown that Zhang Lojiao was a member of the Li Trigrams, and his fellow boxing association member Liu Yulong was also a member of the Golden Bell.[56] Thus we can prove a connection between the Golden Bell and the Eight Trigrams that extends back into the late eighteenth century.

Based on the overall evidence from official sources, we can conclude that these boxing associations had some clandestine links with the White Sun, the Eight Trigrams, and other religious sects. These connections were not broken even after the failure of the 1813 Tianli sect uprising. Zhang Lin in his 1814 confession stated: "Following several years of famine, in the Shandong-Zhili border area heterodox sects have provoked the ignorant folk. . . . Chief among these are the boxing sects (*quanjiao*) such as the Plum Flower Boxers, the Superman Boxers (Yihuoquan), the Two Wolves Boxers (Erlangquan), the Big Red Boxers (Dahongquan), and the Golden Dragon-Lantern Boxers (Jinlongzhaoquan)."[57] That same year the senior president of the Censorate, Qingpu, pointed out in a memorial that "By punishing the boxing sect teachers, the offshoots of the White Lotus can be destroyed."[58]

Of course there existed boxing associations that had no connections with the religious sects. For example, Tang Huanle, also known as "Beaded" Tang, had been the teacher of Feng Keshan, a leader of the 1813 Tianli sect uprising. Tang had taught Feng the techniques of Plum Flower Boxing before Feng joined the Eight Trigrams. According to Tang's own confession, "Three years ago, I learned he [Feng Keshan] had joined the Eight Trigrams sect, and told him he could not be counted among my students if he was a member of a religious sect. I was his boxing master, but not his religious master. You can verify this from Feng Keshan himself."[59] This is additional proof that boxing groups cannot be considered part of a general system involving the Eight Trigrams and deriving from the White Lotus. When the two kinds of groups joined together, then the boxing groups would become an accessory of the religious sect. Yet if it remained independent, these secret societies were simply martial training groups.

The complex relations between boxing groups and religious sects have been described by the Manchu official Nayancheng, who was serving as governor-general of Zhili and imperial commissioner for suppression of the Tianli sect rebels in Zhili, Shandong, and Henan from 1813 to 1815. In his memorials he used special terms to distinguish between groups of plain martial arts followers and those in which boxing was mixed with religious sectarianism. Lao Naixuan in 1898 adopted only that part of Na Yancheng's argument referring to the "Harmony Boxers of the Li Trigrams" and thus mistakenly concluded that the Harmony Boxers were part of the Li Trigrams. Lao Naixuan went further to argue that the Boxers (Yihetuan) also were part of the Li Trigrams. In truth only certain boxing groups were linked to religious sects. The Qing administration recognized these distinctions by establishing the different categories of "heterodox associations" and "heterodox religious sects." Lao Naixuan did not fully understand the complex relations possible between these two kinds of illegal organizations and consequently believed that Harmony Boxers were an adjunct to the Li Trigrams. This view has been perpetuated, even though it is mistaken.

Around the time of the Tianli sect uprising of 1813, the Zhen and the Li Trigram elements of the Eight Trigrams were the most active. The Zhen Trigram was found in Henan around Hua county, while the Li Trigram was widespread in both Shandong and Zhili. At that time the Harmony Boxers, the Red Boxers, and the Golden Bell groups all had links with the Li Trigram and should be classified as its adjuncts, but the Boxers of the post 1895 period were different. The Golden Bell of the late nineteenth century, who had the most influence on the Boxers, were connected with the Kan Trigram, whereas the Li Trigram had entered a decline. The Harmony Boxers of Shibacun exclave of Guan County had no links with religious sectarians at first. Only after 1900 did they link up with Kan Trigram elements. Thus it is correct to say that in 1900 and afterward the Harmony Boxers did become a wing of the Kan Trigram, but not before.

In considering the relationship between the White Lotus sects and the Boxers from the eighteenth century onward, these conclusions emerge: (1) the Boxers of 1899-1900 constituted a traditional peasant movement in that their organization combined elements derived from both secret societies and heterodox religious sects; (2) the distinguishing practices of the Boxers of 1899 were derived from the Golden Bell and Spirit Boxers traditions and were not a legacy from the Harmony Boxers of the pre-1820 period.

Some scholars have argued that there are instances prior to 1820 in which boxing associations and Eight Trigrams sects were enemies. My own research into this question has not turned up any such occurrences, but Dai Xuanzhi in his work suggests that during the 1813 Tianli sect uprisings, the Harmony Boxers and the Eight Trigrams groups were enemies.[60] Recently Fang Shiming has made this same point, based on what the Magistrate Wu Kai wrote in his account *Jinxiang jishi* (Records of Jinxiang County [Shandong]), composed at the time of the 1813 uprising.[61]

Wu Kai claimed that the prisoners Wang Puren and Zhang Heng, both of whom were known members of a Li trigram sect, were also Harmony Boxers. Moreover, on the basis of Zhang Heng's confession, Wu Kai went further to suggest that the two men were enemies of Chui Shijun, also a member of the Li Trigram sect. Wu Kai wrote this in a memorial to the Shandong Governor-General Tongxing on September 13, 1813 (Jiaqing 18/8/19); however, Wu's conclusion is incorrect. When Wu Kai states that the two prisoners were connected to the Harmony Boxers he employs conditional wording, indicating that he lacks conclusive evidence. Thus he writes that Zhang Heng, "seems to be a Harmony Boxer" and that Wang Puren "also may be a stalwart of the Harmony Boxers" while stating that "even the reports of the enmity toward Cui Shijun cannot be completely proven." So all of this lacks certainty. From Cui Shijun's confession we can see that Wang Puren was the leader of the Li Trigrams in Jinxiang, that Wang had brought the Li Trigram teachings into that area, and that he had recruited Zhang Heng, Wang Jingxiu, and Liu Yan as disciples. Finally, Liu Yan recruited Cui Shijan into the Li Trigrams in 1804 (Jiaqing 9).[62]

Wang Puren's own statement says that he originally was a member of the Li Trigrams but in 1812 took the leader of the Zhen Trigrams, Xu Anguo, as his master, thereby leaving the Li for the Zhen Trigrams.[63] Moreover, there is no mention in either Wang Puren's or Zhang Hen's confession that they ever joined the Harmony Boxers. Furthermore, if we look at Xu Anguo's confession we find that Cui Shijan also was Xu's disciple! If Wang and Cui had the same teacher, how could that become a source of contention between them? Thus, I have decided that Wu Kai's conclusions about these matters only conjectures.

Xu Anguo's Zhen Trigram sect operated in the area around Jinxiang, Dingtao, Chengwu, Cao, and Shan counties, and the Harmony Boxers could not have existed in that area if they had been enemies of the Zhen Trigram followers. The Harmony

Boxers in the Jinxiang region were under proscription by the Qing dynasty. Wu Kai himself states, "Although Harmony Boxers were strictly forbidden, they did exist in my district."[64] We find, however, that Wu Kai's policies toward the Harmony Boxers were different from those of other officials; he was lenient toward the boxing groups, while he rigorously suppressed the sectarians.[65] In the face of this policy designed to split the boxing groups from the sectarians, it is not surprising that some individuals took advantage of the situation and in the process made former sectarian comrades into enemies. This is probably the case in the matter involving Cui Shijun, who was found guilty of killing two men who were Boxers but had also become constables. My conclusion is that around 1813 the boxing associations had become adjuncts or wings of the religious groups, and it is highly unlikely that these two allied types of organizations ever became enemies without official connivance such as Wu Kai employed.

Fang Shiming suggests that similar hostility existed between the White Lotus and the Boxers at the end of the nineteenth century.[66] He believes that the White Lotus worshipped the Maitreya Buddha while the Boxers revered the Sakyamuni Buddha, and that became the source of their enmity. This is probably wrong, for Boxer placards and writings do not contain praise for Shakyamuni rather than Maitreya. Fang Shiming quotes a passage from the text, "Yuan Li beiwen" (Yuan Li Inscriptions) found in Jinghai County that states, "In the one-thousand nine-hundred and ninety-third year, Sakyamuni leaves the world and it will be transformed," while the prophetic writing "Gengzi pingfenglu" (The Disaster of the Genzi Year [1900]), changes the word "leaves" to the word "enters." The difference in this one word alters the meaning, and Fang Shiming believes that the concept of Sakyamuni "entering" the world at the end of the present era became widespread. His interpretation would lead toward increased devotion toward Sakyamuni Buddha as a Buddha connected with a future era, whereas the original wording would maintain the standard Buddhist doctrine that the Maitreya Buddha is the Buddha of the future world.

According to the pre-1820 work "Sanfo ying jiejing" (The

Text of the Kalpas of the Three Buddhas), "We believe that in the past was the era of the Lamplighter Buddha, that the present is the era of the Sakyamuni Buddha, and in the future Maitreya Buddha will come."[67] Also, the leader of the Return to the Origin sect (Shouyuanjiao), Fang Rongsheng, confessed that he had become a disciple of Jin Zongyou in 1808 (Jiaqing 13) and that

> Jin taught from a text which described the Lamplighter Buddha seated on a green lotus, representing the unparalleled, and teaching the "Green Sun doctrine"; while the Sakyamuni Buddha sat on a red lotus, represented the great ultimate, and taught the "Red Sun doctrine"; and finally, the Maitreya Buddha sat on a white lotus, represented the supreme ultimate, and taught the "White Sun doctrine."
>
> Today the "great ultimate" [representing Sakyamuni] is withdrawing, indicating the approach of the Maitreya Buddha's era of the supreme ultimate. This teaching disturbed the people.[68]

According to the beliefs of the White Lotus followers, history passes through three stages or kalpas; at each stage a Buddha appears to save the world for the next. "The Eternal Mother bore 9,600,000,000 daughters and sons. In the previous stage of the Green Sun doctrine, the Lamplighter Buddha saved 200,000,000 souls, while the Sakyamuni Buddha brought 200,000,000 souls with him into this kalpa of the Red Sun doctrine. For the White Sun kalpa the Maitreya Buddha will try to save 1,000,000,000."[69] From these texts we can see that the White Lotus ideas do not put Sakyamuni and Maitreya into competition, but rather take the Maitreya Buddha as the truest and final Buddha. Fang Shiming's interpretation of the text in "Yuan Li Inscriptions" probably is without basis.

Fang Shiming also quotes from a text, "Nanyuan santaishan shenzi bei" (Inscriptions Concerning the Spirit of Three-Tier Mountain at Nanyuan). The text reads, "From the West come those who study the Maitreya Buddha; while those from the North despise the Buddha and are uneasy about other deities. They put the Jade Emperor at the head of all deities."[70] Fang misreads this text by interpreting the initial part of this passage to read, "From the West come those who study. The Maitreya Buddha is the evil Buddha from the North." This is a clear distortion of the grammar of the text. Thus neither of Fang's quotations proves the opposition between the Boxers and the White Lotus at all. On the contrary, when correctly read, both texts indicate that there existed a White Lotus influence upon the boxing associations.

Dai Xuanzhi also believes the White Lotus and the Boxers were in opposition. He says the White Lotus are distinguished by the eight-character chant "zhenkong jiaxiang wusheng fumu" (Our Homeland of True Emptiness; Eternal Mother). He argues that the Boxers never used this formula. Also he argues that the White Lotus worshipped only the Maitreya Buddha whereas the Boxers honored the Jade Emperor, the Eternal God of War (*wusheng guandi*), and other figures taken from stories of martial adepts and spirits.[71] The formula of the White Lotus that Dai Xuanzhi quotes, however, dates from the White Lotus uprising of the eighteenth century. By the time of the late nineteenth century, the historical conditions had changed and the chant "Our Homeland of True Emptiness; Eternal Mother" was no longer appropriate; so it was replaced by others. The change in the deities revered by the Boxers also reflects alterations to meet different historical circumstances.

The broadening of the Eight Trigrams' pantheon of worship from Maitreya to the Jade Emperor, the God of War, and even to characters drawn from popular novels and opera probably dates from the Daoguang reign (1821-1851) when such practices began appearing in Zhili around Julu and Qinghe counties. When we look at the changes in the propagation of the Eight Trigrams sect during the nineteenth century, we find that in 1823 (Daoguang 3) the Li Trigram leader Li Fangchun had confessed: "In propagating our faith we often meet with the ignorance of common people and so we assumed the identities of characters from the novel *Fengshenyanyi* (Canonization of the Gods). Huang Dangxuan became the incarnation of Nazha, and others did likewise to show those who had died already still had another existence to come and were not doomed forever."[72] Huang Yubian in his book *Poxie xiangbian* (A Detailed Refutation of Heresies) includes sixty-eight White Lotus scriptures including two entitled "Sutra Concerning Saving the State, Protecting the People, and Subduing Demons" and "Sutra of the Three Virtues: Protecting the Country, Saving the People, and Subduing Demons." The first describes Daoist Gods who protect the Jade Emperor and various stories connected with the God of War in which he changes everyone into a saint or saves the world. The latter scripture takes several episodes from the *Romance of the Three Kingdoms* and relates how figures from those stories loyal to God of War were saved.[73]

According to Huang Yubian, who served as local magistrate in these areas of Zhili in the 1840s and 1850s, these sutras were discovered in the region around Julu and Qinghe, which is adjacent to the Grand Canal and Shandong province. The previously mentioned Li Fangchun, leader of the Li Trigrams who confessed in 1823, hailed from the Qinghe region.[74] From his account and other evidence, we can conclude that after the failure of the 1813 Tianli sect uprising, the Eight Trigrams had changed the gods they worshipped and altered their religious teachings during the Daoguang reign. We must take account of those changes and cannot imagine that the religious ideas of the pre-1813 period were transmitted unaltered down to the 1890s. In fact, there is evidence that Zhu Hongdeng's activities may represent a derivation of the Li Trigram teachings developed by Li Fangchun.

The foregoing discussion proves that some relationship did exist between the late-nineteenth-century Boxers and the pre-1813 White Lotus. In the initial stages of the Boxer movement the contributions of the sectarians were revealed quite clearly, but as the movement matured these sectarian influences became less obvious. Finally, when the Boxer movement reached its peak, we cannot locate any special sectarian activities. This is because the organizational characteristics of primitive secret societies and religious sects could not meet the requirements of a true anti-imperialist struggle.

Still, this does not mean that the influence of sectarianism had disappeared completely from the Boxer movement. Upon close examination it can be seen that most sectarian influences in the Boxer movement came through the participation of Buddhist monks and Daoist priests. Both documentary sources and oral history interviews prove that these monks and priests were key figures in the Boxer struggle, and that most of them were members of the Eight Trigrams.

In 1896 when the Big Swords were active in Shadong's Cao and Shan counties there was a Daoist priest of whom it is recorded, "Wang lives south of the county seat of Shan and is leader of the Big Swords. He also is an adept in the martial arts and he has a large group of followers who believe that he can administer an incantation that makes them invulnerable to injury. When officials and troops tried to arrest him, he disappeared without a trace, but it is said his followers now have spread among villages throughout the region."[75] Since Shan county's Big Swords were part of the Kan Trigram network, priest Wang also must have been part of that network.

In the case of Zhu Hongdeng's activities in Pingyuan county, there were many monks and priests involved through Monk Benming, the boxing master. According to records about the engagement at Gangzi Lizhuang, among Zhu Hongdeng's Spirit Boxer ranks were "monks and priests. . . . His ranks were mostly outsiders, including former militiamen, and so they knew how to fight."[76] Thus, sect members were key figures among the Boxers in Pingyuan county also.

The Monk Xufu, a Boxer leader in Chiping, had connections with Benming and Zhu Hongdeng. He had started out at the Mingjiao Temple at Sanlitang and had been active for many years around Chiping and its neighboring counties of Boping and Changqing. The Qing authorities had labeled him a "concealed evil-doer."[77] Among the key members of Yan Shuqin's Harmony Boxers there was a monk surnamed Dong.[78] In Zhang Sendou's Harmony Boxers there were three Daoist priests, Zhu Jiubin, Liu Hualong, and the blind Han Er. All of them were sectarians devoted to the overthrow of the Qing and the restoration of the Ming.[79] In Zhili at Jingzhou, the Monk Wu Xiu was a boxing leader and "a follower of the Eight Trigrams."[80] At Gucheng the boxing leader, the Monk Dagui, was "a follower of the Li Trigram." At Zhuozhou the leader of Boxers was the Monk Longxiang.[81] In Beijing there was the Daoist priest Chuanxin,[82] who was proclaimed a "living Buddha," and in addition there was a Grand Master, the Monk Daotong, and another Daoist priest called "Master of the Hall of Enlightenment."[83] All of them were members of heterodox sects before becoming top leaders of the Boxers.

The White Lotus tradition may provide an explanation of why so many of these monks and priests became leaders in the Boxer movement. The White Lotus had absorbed much from Buddhism and Daoism in its genesis and development. After 1813 the transmission and teaching of the White Lotus doctrines was conducted by monks and priests in many locations. For example, the eighth patriarch of the Clear Tea sect, Wang Bingheng, recruited many monks and Daoists to help propagate his teaching. From that time onward the White Lotus tradition operated in this fashion and thus came to color the Boxer movement from the start to finish.

The Boxers did not originate from the militia

Another important issue is the argument that the Boxers may have originated from the militia. George Nye Steiger was the first to propose this theory,[84] and Dai Xuanzhi expanded upon Steiger's ideas.[85] Steiger emphasizes an edict dated November 5, 1898 (Guangxu 24/9/22), calling for the organization of militia in Zhili, Shandong, and Fengtian provinces. He believes that the Boxers started from this. Aside from this he does not produce any convincing evidence to support his interpretation.

Dai Xuanzhi's main evidence comes from Shandong Governor Zhang Rumei's letter to the Zongli Yamen of June 17, 1898 (Guangxu 24/4/29), and his memorial to the emperor dated June 30 (Guangxu 24/5/12). That memorial states:

In the administrative regions along the border between Zhili and Shandong, the people are fond of boxing practices, and these groups have set up as unofficial militia (*xiangtuan*). They were first known as the Harmony Boxers and later changed their name to the Plum Flower Boxers. Recently the name Harmony has reappeared. These people have always claimed that the word "Yihe" in their name meant "good" or "righteous." They claim to be a newly established group, but still they predate the appearance of Christian churches in this area during the 1850s. So they are not anti-Christian, but rather attempt to protect their families and defend against banditry.[86]

This evidence seems to make the militia origins theory difficult to refute. In fact, the passage contains several elements that are difficult to interpret.

First, the memorial quoted above contains Zhang Rumei's explanations and was based on a firsthand investigation in Guan county ordered by Zhang Rumei and made by Li En-xiang, the Jining prefect; Hong Yongzhou, the Dongchang prefect; and Cao Ti, the Guan county magistrate. Zhang's explanations remain somewhat doubtful, and even though I have never seen the full reports of these three officials, we can see incorporated in Zhang's memorial to the throne the outlines of their original reports. Hong Yongzhou's report to Zhang Rumei was dated June 8, 1898 (Guangxu 24/4/20). It stated:

Plum Flower Boxers, originally known as the Harmony Boxers, flourished in the Shandong and Zhili border region. The people in those areas are fond of martial arts, and even when residing peacefully many become highly practiced in these martial skills for mutual protection and self-defense. The teaching of boxing was widespread, and its practice became quite common in Henan, Shanxi, and Jiangsu, with various styles vying for popularity.

Every spring at the local fairs groups of boxing adepts take advantage of the occasion to put on demonstrations called "boxing exhibitions" (*liangquan*) before the crowds. Among the common people these events were also called Plum Flower Boxing Meets.

Following last year's disturbances over religious matters at Liyuantun, this spring Plum Flower Boxing became a mass movement and so an official order for disbandment was issued. Since that time, these [boxing adepts] have gone back to the old name "Harmony Boxers" and have dropped the name "Plum Flower Boxing" when teaching or displaying their skills among the people.[87]

This account more clearly reveals that the groups passed through a cycle in which they began as Harmony Boxers, then became Plum Flower Boxers, and finally resumed the exact appellation of "Harmony Boxers." Zhang Rumei, in his subsequent letter of June 17 to the Zongli Yamen, based his account of the Boxers' history on Hong Yongzhou's report. Zhang's account states the facts accurately and never uses the term "Harmony Militia" (Yihe xiangtuan). Yet thirteen days later, on June 30, in a memorial to the throne, Zhang Rumei changed his terminology so that the names became "Harmony Militia" (Yihetuan), "Plum Flower Boxers," "Harmony Boxers," and he particularly referred to the Boxers as militia.

In this second account, Zhang Rumei states his memorial was based on the field investigations of both Hong Yongzhou and Cao Ti. Yet, when we compare Zhang Rumei's June 30 memorial with materials from Cao Ti's collected works, *Guchuncaotang biji* (Notes from the Studio of Springtime Past), it turns out that Zhang's version is based on Cao Ti's writings and not those of Hong Yongzhou.

Cao Ti claimed that the Plum Flower Boxers and Harmony Boxers had evolved from local militia in order to serve his own purposes. He was the magistrate of Guan county, but Shibacun exclave (literally "Eighteen Villages," but in 1898 containing more than that number) was separated from the county by sixty kilometers and therefore was outside his effective control. Boxing practices already were well established in that area. In 1898 these boxers "had assembled a huge crowd one night in April" at Shibacun and at nearby Shaliuzhai in Wei county.[88] Cao Ti had great difficulty extending his authority into this Shibacun territory and consequently, hoping to ease the danger of his administrative problems, attempted to enroll the boxing groups as local militia (*xiangtuan*). In fact, Cao Ti's approach fitted the policy Zhang Rumei had promulgated on April 28, 1898 (Guangxu 24/3/28), calling for the creation of local militia. Thus we see that Zhang Rumei had abandoned Hong Yongzhou's approach for dealing with the Boxers in hopes of being better able to control the crisis by using Cao Ti's suggestions.

Second, we need to determine if Cao Ti's story is accurate.

According to his own account he went to Liyuantun for about a month, accompanied by a secretary and a couple of constables, sometime in April and May 1898 (Guangxu 24/2-3).[89] On closer examination it seems that Cao Ti's account is false: three instances can show how that is probably the case. In his collected writings, Cao Ti states that there was a grain storehouse at Shibacun where the local taxes were collected. In fact, taxes from Shibacun were collected at the market town of Ganji, and the grain storehouse was located in the Leyu Academy (*shuyuan*) in Ganji. We also know that ever since 1887 the disputes between the Christians and other people in Shibacun had increased. For many years none of the local magistrates had dared to enter Shibacun unprotected and remain overnight.

According to local historian Guo Dongchen's account of these events in 1898, two officials--the Dongchang prefect (Hong Yongzhou) and the Guan county magistrate (Cao Ti)-- and three others went to Ganji market town and called a meeting of the gentry and influential people of the area, including the Boxer leader, Zhao Sanduo. The purpose of the meeting was to find a settlement for the dispute over the Christians' construction of a church on temple grounds.[90] This visit included Cao Ti; he accompanied the others to Ganji and did not go alone to Liyuantun. The meeting at Ganji was chaired by the intendent of the Donglin Circuit in Zhili. In his own account of these events Cao Ti claims that after his repeated exhortations, Zhao Sanduo finally consented to disperse several thousand Boxers who had gathered at the crossroads in Liyuantun. This is a complete fabrication, for in our interviews with both ordinary people and former boxers at Liyuantun, none of them could recall any such incident. Cao Ti, then, has colored his recollections in his collected works to make himself out as a hero. Consequently, I have concluded that Cao Ti's accounts and his reports to Zhang Rumei are untrustworthy, and his evidence about the Boxers must be considered suspect.

Third, the relationship around Shibacun exclave among the local militia, the Boxers, and the Christians had become quite complex. In these areas, especially at Houweicun and Shaliuzhai of Wei county in Zhili, the Red Boxers, the Plum Flower Boxers, and the Harmony Boxers all had their own history and traditions extending from the eighteenth century. All three groups of boxers were completely separate from the local militia. The militia dates only from the period from the 1840s through the 1860s. We do not know even today if the Shibacun exclave had its own militia in those years, but in the adjoining Wei County a militia was established in the summer of 1865 (Tongzhi 4). At that time Expectant Censor Cheng Tingjing requested the throne to appoint some able person to train a militia in southern Zhili. Zheng Yuanshan, a former governor of Henan, was residing at Guangzong, and under his careful direction local militia units were established in the region. In 1870 (Tongzhi 9) after the defeat of the Nian Army, the local militia had fulfilled its purpose and was disbanded. Around 1894 (Guangxu 20) the Wei county magistrate, Zhang Lianen, reconstituted the militia for the purpose of defending against banditry. In 1901 (Guangxu 27) this militia was disbanded a second time "because of a fear that the militia might become involved in the disturbances."[91] All these facts are recorded in the local gazetteer.

The local historian Guo Dongchen contributes some additional information. The Wei County militia established in the Tongzhi reign (1862-1875) was divided into three sections-- those from south of the county seat were called the "Comity Militia" (Zhihetuan), those from the east were the "Concord Militia" (Peiyituan), and those from the north were the "Harmony Militia" (Yihetuan). The head of the Harmony Militia was Zhao Laoguang, a literati from Beitaiji. His militia was disbanded in 1870 (Tongzhi 9), but reconstituted in 1896 (Guangxu 22). In 1896 Zhao Laoguang was still alive, and once again he became head of the Harmony Militia in the territory north of the Wei county seat. In November 1898 (Guangxu 24/9) when Zhao Sandou led the Harmony Boxers in an uprising, the Catholic converts demanded that the militia be brought out to defend them.

> The militia commanders and village militia leaders discussed [the Catholic demand] at a meeting and concluded that Zhao Sanduo was neither in rebellion nor a bandit. Moreover, the dispute with the Catholics fell into territory under jurisdic-

tion of Shandong and was also a private matter in which the militia should not intervene. The militia also [according to regulations] could not be called out prior to the outbreak of disturbances. In addition, many of the militia members were practitioneers of Harmony Boxing, and so, fearing strife within the militia, Zhao Laoguang decided not to comply with the Catholics' demand.[92]

Thus it is clear that in 1898 the official militia of Wei county and certainly the Harmony Militia from the northern part of the county were not opposed to the Harmony Boxers. Still these two bodies were not exactly the same, for the boxing associations maintained an identity separate from the militia. There were connections between the two, but their differences are noteworthy, especially those in the purpose and organization of the two bodies. In particular, Zhao Sandou was the leader of the Plum Flower Boxers and not a militia commander, so he could not have led the Harmony Militia in an uprising.

The actual situation in the northern part of Wei County is even more complex. When Zhao Laoguang would not agree to answer the demands of the Catholics for militia protection, they then brought together people from several villages, including Weicun, Panzhuang, Zhaozhuang, Chenzhuang, Houlizhuang, and Xizhuguanying, into another militia-type body called the "Harmony Association" (Yihehui) to resist Zhang Sanduo.[93] These Christians were armed and established their militia on November 1, 1898 (Guangxu/9/18). According to the diary of a Catholic priest, "Our muster roll contains 166 men from Zhaojiazhuang, all of whom are armed; 196 from Weicun, and 117 from Panzhuang, Chenjiazhuang, and four other villages, to make a total of 477 [sic] men."[94]

Thus in northern Wei county there were three separate types of martial organizations--a boxing association, an official militia, and a private Catholic militia--all of which employed the term *yihe* in their names. The official militia was neutral in the local dispute, but the Harmony Boxers and the Harmony Catholic militia were in direct opposition. When Zhang Rumei mentioned in his memorial of June 30 about a local militia with the name Harmony that had been established in the Xianfeng

(1851-1862) and Tongzhi (1862-1875) reigns, he was referring to the local militia operating around Houweicun and Shaliuzhai in northern Wei county, only about three kilometers away from Shibacun, even though these two territories were under the authority of different provinces. The Shandong officials, because they were so far away from Shibacun, were unable to carry on inquiries on the scene to determine accurately any connections between the Harmony Militia and the Harmony Boxers and had to rely upon hearsay evidence for their conclusions.

Zhang Rumei's account that the Boxers developed from a militia has been picked up by several historians. Yet, that explanation of the Boxers' origin from a militia is wrong, just as was Lao Naixuan's explanation of their origin from the Li Trigrams. Zhang Rumei's version provided a basis for those officials who advocated a policy of lenient pacification, while Lao Naixuan's theories gave support for those officials who wanted to suppress the Boxers using all necessary force. Both of these incorrect explanations have been passed on to later generations and have confused a great many specialists in Boxers studies. Some specialists have become entangled in endless disputes that derive from these two incorrect explanations.

The first occurrence of the term "Yihetuan" for a local or an official militia was not the case in Zhili's Wei County. Records show that in 1862 an official militia with the name "Yihetuan" was established in Shandong's Yanggu county on May 11 (Tongzhi 2/4/13). Its purpose was to defend the county and resist attacks from a White Lotus army then operating in northwestern Shandong.[95]

Even during the period of the Boxer movement of 1899-1900 there were places where the official militia used the name "Yihetuan." For example in Shandong's Shouzhang county there was a Harmony Militia which had a strongly reactionary character. The officials and gentry in this region established a number of militia units in order to suppress the Big Sword Society. On both sides of the Yellow River there were militia units with names such as "Benevolence" (*renhe*), "Perfect Benevolence" (*tongren*), "Joint Harmony" (*yihe*), "Virtue" (*renyi*), and "Cooperation" (*xiehe*). There were four militia units north of the Yellow River and four to the south that all used the name "Joint Harmony" (*yihe*), a name that uses one different character but is homophonous with the more familiar "Yihe."[96] These militia units in Shouchang absolutely were not part of the Boxer movement, because in this area the militia and the boxing groups were in opposition all during these years. Proof of this point can be found repeatedly in the collections *Shandong Yihetuan anjuan* (Documents on the Boxer in Shandong [1980]) and *Shandong Yihetuan diaocha ziliao xuanbian* (Selected Survey Materials on the Boxers in Shandong [1980]) and need not be presented here again.

How did the Harmony Boxing Association become the Boxers?

We already have clarified how the Boxers did not originate from an official militia, but the problem remains of how these various boxing associations--including the Harmony Boxers, the Big Swords, the Spirit Boxers, and others--were transformed into the Boxers (Yihetuan). The significant change here is one from a boxing group (*quan*) into a militia (*tuan*), which appears to be a transmutation from a basically illegal heterodox association (*xiehui*) into a legal and official militia, or at least into a semi-legal form of organization for self-defense. These are major issues in the history of the Boxer movement that have been disputed for many years. In this section, I present my own conclusions on these matters.

The Harmony Boxers, as already discussed, appeared in Guan county in 1898 (Guangxu 24) after changing their name from Plum Flower Boxers. They became active in March and April 1898, but because of Governor Zhang Rumei's policy of suppressing information about them most people in China and overseas had not heard about their activities. Prior to October 1899 even the name "Yihequan," most commonly translated "Righteous and Harmonious Fists," appeared only occasionally in the diary of a French Catholic father or in the telegram of a British Protestant missionary. The press, both Chinese and foreign, referred to the Big Swords and the Spirit Boxers but made little mention of the "Righteous and Harmonious Fists."

The only official who discussed them was Lao Naixuan, stationed at Wuqiao county in Zhili. He published *Yihequan jiaomen yuanliu kao* (On the Origins of the Boxer Sectarians) in October 1899 in which he announced the "Yihequan" were an offshoot of the Li Trigrams.

Widespread usage of the name "Yihequan" dates from the incident at Gangzi Lizhuang on October 21, 1899 (Guangxu 25/9/7). The establishment of boxing grounds within Pingyuan and En counties dates from April and May 1899. By September 1899 their activities had attracted official notice, requiring Shandong Governor Yuxian to send a telegram to the Governor-General of Zhili, Yulu, about these boxing groups, but he did not refer to them as "Yihequan."[97]

At that time the foreign observer closest to the scene was the American missionary H. D. Porter, who lived at Panzhuang in En county. In his letters he mentions the Big Swords, the Plum Flower Boxers, and the Spirit Boxers, but not the "Yihequan." This provides evidence that the Harmony Boxers were a secret association and were not a private militia for self-defense, and certainly not an official militia.

After the incident at Gangzi Lizhuang in October 1899 a boxing adept from that village, Li Changshui, went to the Liuli Temple in Gaotang to request that Zhu Hongdeng and his Spirit Boxers come to help.[98] At this juncture the government still opposed the movement, and this is proven by the engagement at the Senluo Temple on October 18 (Guangxu 25/9/14) in which the Harmony Boxers fought with the Qing troops. This is called the Pingyuan incident, and only after it did the use of the word "Yihequan" become common.

There is documentary evidence to support this conclusion. On October 14, 1899 (Guangxu 25/9/10), H. D. Porter at Panzhuang wrote in a letter:

> It is just a month this morning since we first secured a few soldiers from the District city to defend us against possible, but seemingly imminent, attack upon us by turbulent members of the Yi Ho Chuan Society. The last character in the combination of three in the name of the society is "Fist," or "Boxers," and the Society has been named for the past year

"Spirit Boxers," a company of young fellows who have gathered together for wrestling and general gymnastics, with the underlying purpose of combining against all foreigners within range. Under the pretense of a patriotic purpose, "Exalting the Dynasty and Destroying the Foreigner," these companies have increased in great numbers until the whole region about us was wholly infiltrated with them.[99]

Only after this letter, written three days after the Pingyuan incident, did the term "Yihequan" begin to appear in the Chinese and Western press.

According to the official archives, there were references to the "Yihetuan" two weeks later at the battle of the Senluo Temple. The local villagers from Dazhifangcun where, the temple was located, refer to the incident as the "clash between the official troops (*guanbing*) and the militiamen (*tuanmin*)." A petition written by a literatus of that village, Pei Xiuting, states: "On the 14th of last month [October 28, 1899] the official troops and the militiamen clashed at the outskirts of the village. The militia were defeated and withdrew northward. They escaped by dispersing."[100] This marks the beginning of the switch from the name "Yihequan" to that of "Yihetuan."

Did the new Shandong governor, Yuxian, instigate this usage of the term "tuan" to designate the local unit at Gangzi Lizhuang? It is widely believed that he did. Actually, the evidence is insufficient to support such a conclusion.

In 1898 when Yuxian served as Shandong provincial judge, he supported fully Zhang Rumei's policies of appeasement. Governor Zhang had advocated changing boxing associations into militia and transforming private units into public ones. However, when Yuxian became governor in April 1899 his policies changed. Yuxian understood the long-standing difficulties between the Christian converts and other Chinese peasants. He had agreed with the earlier moderate "policy of disbandment," but since those approaches had not worked, he changed to a harsher policy of suppression when he became governor.[101] Following his elevation to the Shandong governorship in April 1899, Yuxian stated, "More than eight times in eight months [I have] forbidden the people to establish groups such as the Big Swords and Spirit Boxers, and further prohibited the public teaching of boxing in order to avoid any disturbances arising from large crowds."[102]

After the Pingyuan incident of October 28, 1899, Yuxian felt compelled to relieve Jiang Kai of his duties as magistrate there because Jiang Kai's stupidity had led to the death of innocent people. It was this reason, rather than any bad feelings between Jiang Kai and Yuxian, that led to Jiang Kai's replacement.[103]

Upon close examination of the imperial decree of November 20, 1899 (Guangxu 25/10/18), we can see that the Qing government concluded that Yuxian had covered up for Yuan Shituan, the troop commander in the Pingyuan incident. The Court felt Yuxian had been too lenient, for Yuan Shidun simply had been reassigned to serve in the army of his brother Yuan Shikai. The decree stated: "Clearly you [Yuxian] have been too lenient in handling this matter. Is it proper for an official with such a high post [as governor] to act in this manner?"[104] Yuxian's policy following the Pingyuan affair had been "capture the leadership and disperse the rest." It is possible that "dispersal" may have permitted some men to rejoin a militia group, but nonetheless Yuxian's policies were more harsh than Zhang Rumei's. The latter advocated turning boxing associations into official militia.

The accounts of both the Gangzi Lizhuang and the Senluo Temple incidents indicate that the Boxers themselves began applying the term "Yihetuan" to their organization. Newspaper accounts carry the same implication:

The Yihetuan bandits in Shandong rashly spread rumors agitating the common people. Their power grows and their influence spreads each day. . . . Fearing official suppression they began calling themselves a militia to indicate they were established for self-defensive purposes. Fearing a loss of popular trust, they have announced their support for the Qing along with their intent to destroy the foreigners. This falsely claimed talisman is being used to cover up their criminal actions and dupe the people.[105]

Although the Yihequan has started to be referred to as if it were a militia, it really is a branch of the Eight Trigrams. Now they refer to themselves as the Yihetuan and in the name of anti-Christianity tyrannize the people.[106]

This new title, "Yihetuan," spread throughout the area around Pingyuan and Chiping. Older people who themselves witnessed the battle at the Senluo Temple have recalled, "The "Yihequan" was also called the 'Yihetuan' after the battle at the Senluo Temple. They were called a *tuan* because more than 500 followers of the Yihequan assembled here and formed into units (*tuan*). We don't know exactly how many men were in each unit. The meaning of the term "Yihe" was a common, just purpose."[107]

Similar explanations were given in the Zhangqing and Gaotang regions. Yet in those regions, the Spirit Boxers had to change their name into the Yihetuan. Old people from that area recalled the change in this manner: "All the villages set up their own boxing grounds where everyone would gather. When a great assembly took place on these grounds it was called a *tuan*."[108]

In Pingyuan county the explanation given for why the "Yihequan" became known as the "Yihetuan" is that "They had many groups and each group had its own grounds. Each ground had its own Grand Master and his assistant, or Junior Master. The ordinary members referred to each other as elder and younger master. Each unit (*tuan*) had twenty-four members."[109]

Still, nowhere--not in Chiping, Gaotang, or Pingyuan--were these Boxers a true official militia. The name "Yihetuan" was self-proclaimed and spread among the branches of the "Yihequan." It was the end result of the increasing cohesion among the various boxing associations known as the Big Swords, Spirit Boxers, and Harmony Boxers.

Prior to 1900 around Pingyuan and Gaotang there were two different varieties of Yihetuan. Local stories distinguish between groups "with roots" and those "without roots." The people at Ershengtangcun in Pingyuan recalled, "Our Yihequan had no roots, and thus was independently established. We did not send our Boxers to participate at the battle of the Senluo

Temple."[110] What the term "without roots" means here is that it was an independently established unit without any connection to Zhu Hongdeng's Spirit Boxers or any other boxing association. It does not mean a privately established irregular militia.

Of course while this wave of antiforeignism was increasing, there were some landlord and opportunist elements who did joint or establish their own private groups, but these were the exception. Most of these self-proclaimed Boxers were composed mainly of poor peasants, no matter if they had links with other boxing groups or not. The Boxer movement certainly resulted from a broadening peasant struggle. Once the term "Yihetuan" appeared after the Pingyuan incident, it was used in official correspondence, newspapers, and memorials with different meanings. Some officials felt that Yihetuan were good people and should be supported; others felt they were a kind of outlaw who must be suppressed.

Until January 11, 1900 (Guangxu 25/12/11), when the Qing official policy was announced, there had been no unified government policy toward the Boxers. Both the policies of lenient pacification and harsh suppression had been employed by provincial and local officials and were not the Qing Court policy. These local reports and memorials dealing with the Boxers at best only showed the imperial notation "Duly Informed" (*zhidaole*), a set phrase indicating the emperor was aware of the action taken and had assented. The January 11 pronouncement changed the dynasty's long-established policy of suppression of any group with heterodox connections. Previously, whenever a religious sect or secret society appeared, it was labeled heterodox and not allowed to exist openly or legally. Now, on January 11, 1900, Court policy changed with an indirectly worded statement: "Because We have understood there are differences among associations, We need to know if they are criminal or rebellious. If not, it is of no matter if they are sectarian or societies." This new stance obviously gave the Court a means for harnessing the Boxers' increasing strength. The foreign legations in Beijing offered loud and prolonged opposition to this new policy. Yet it must be pointed out that this change of the Qing Court's attitude toward the Boxers was not primarily an attempt to use the Boxers, but had been forced upon the Court by the growing strength of the Boxer movement.

The Qing Court never recognized the Boxers as an official militia. In fact, even after this January 11 document, the Court continued to refer to them as a "boxing association" (*quanhui*) for a considerable period. Also they referred to the Boxers as a militia-like association (*tuanhui*) to distinguish them from an official militia.

On June 2, 1900 (Guangxu 26/5/6), the boxing adherents occupied the city of Zhuozhou in Zhili and raised a banner that proclaimed themselves the "Yihetuan." This was an important stage in the development of the Boxers; yet even at Zhuozhou this appellation was self-proclaimed and was not bestowed on them by the Court. Four days after the Boxers occupied Zhuozhou the Imperial Court proclaimed on June 6, 1900 (Guangxu 26/5/10), "The Yihequan association established in the Jiaqing period [1796-1820] was proscribed, and recently there have been followers of boxing practices who have established militia-like associations (*tuanhui*) for the purpose of opposing sectarianism. They are completely unacceptable for they wrongly equate their interests with those of the Qing state. . . . These boxing adherents ought to disband respectfully and return to their ordinary existence."[111] These remarks were intended to include the Boxers assembled at Zhuozhou.

On June 22, 1900 (Guangxu 25/5/15), Manchu Salt Censor Wenli wrote in his ten-item essay "Proposals for Dealing with the Boxing Associations" a suggestion that the Boxers be declared some kind of a militia and then they could be ordered to disband.[112]

Some say that Zhao Shuqiao's memorial of June 10 (5/3) and Gangyi's of June 25 (5/18) both advocated support and use of the Boxers, and that policy indicates that these officials intended for the Yihetuan to become an official militia. Their advocacy did not cause the Qing Court to change its policy. Even up until July 1, 1900 (5/24), Court edicts stated that "neither pacification nor suppression is suitable" or "it is difficult to know how we will be able to resolve this question."[113]

When the imperial declaration of war against the foreign states was made on July 2, 1900 (5/25), the Yihetuan suddenly became "good people" (*yimin*). The Court's endorsement of the Boxers can be seen in the announcement of special honors to be awarded to those provincial governors who established such militia units in their territory. Qing policy had undergone a further change at this point, abandoning the approach that combined pacification and suppression, and adopting one of full support for the Boxers.

Yet this policy of support only lasted a few days before it was changed. On July 7 (5/30) the Qing Court explained the changing policy in an indirect manner to provincial officials: "Suppression [of the Boxers] might put Us on the brink of disaster by feeding the configuration. We need to act carefully to save the situation."[114] The Court's words imply that suppressing the Boxers might produce a general antidynastic uprising in North China. Just four days later, on July 11 (6/4), the Qing Court again announced a return to the policy of "emphasizing harsh suppression."[115]

Thus the Qing Court only endorsed the Boxers as an official militia and "good people" for a few days in July 1900. That policy was a temporizing one brought about by the Court's mistaken initial impression of the Boxers' power. When looking at the Court's policy during the whole history of the Boxer movement, the Court used established dynastic policy in dealing with both the Yihequan as a boxing association and the Yihetuan as a militia-like association. Those policies opposed the existence of such bodies and so help prove that the Boxer movement itself was based on the interests of the peasantry and advocated rural class struggle.

In conclusion, the Boxers' organization and the origins of their groups reveal the influence of the White Lotus and represent a combination of the Eight Trigrams sect with boxing associations. All of these characteristics mark the Boxers as a traditional peasant uprising. The Boxers' activities and their methods of recruitment, especially their religious practices and coloration by superstition, reveal their primitive nature. Yet beneath their veil of superstition and religion we can discover the strong revolutionary spirit embodied in the Boxers. Engels, in analyzing the sixteenth-century German peasant wars, pointed out that there are three sources of peasant movements, "mysticism, open heathenism, and armed insurrection."[116] Of course the conditions that brought about the Boxer movement and the content of their struggle were different from the German peasant wars, but looking at them both as examples of traditional peasant movements we can see that the Boxers share the same three characteristics.

According to Engels's view it is impossible to avoid religious superstition and backward ideas in peasant movements, and thus we must take more into account than a peasant movement's shortcomings in making our evaluations. We must employ a more historical view of the Boxer movement's deficiencies and thus can conclude that although the Boxers were primitive, their organizational forms and concepts were appropriate to what peasants of their day could have achieved acting on their own. Thus those who improperly exaggerate the Boxers' superstitious ideology deviate from the Engels's views in their appraisals.

Notes

1. "Yihetuan yundong chuqi douzheng jieduan de jige wenti," in Zhongguo renmin daxue, Qingshi yanjiuso (Qing History Research Institute, Chinese People's University), eds., Zhongguo jindaishi lunwenji (Selected Articles on Modern Chinese History) (Beijing: Zhonghua shuju, 1979), vol. 2, pp. 661-92. This article appeared originally under the pseudonym of Wu Songling in Shandong daxue xuebao (Bulletin of Shandong University) 2 (1960).
2. Fan Wenlan, Zhongguo jindaishi (Modern Chinese History) (Beijing: Renmin chubanshe, 1953), vol. 1, part 1.
3. See my article, "Yihetuan yundong."
4. Guo Moruo, ed., Zhongguo shigao (Draft History of China) (Beijing: Renmin chubanshe, 1962), vol. 4.
5. Zhen Zhanruo, "Yihetuan de qianshi" (Early History of the Boxers), Wen shi zhe (Literature, History, and Philosophy) 3 (1954).
6. See my article, "Yihetuan yundong."
7. Li Shiyu, "Yihetuan yuanliu shitan" (An Analysis of the Boxer's Origins), Lishi jiaoxue (The Teaching of History) 2 (1979).
8. Qian Bozan, ed., Zhongguo shi gangyao (An Outline of Chinese History) (Beijing: Renmin chubanshe, 1964), vol. 4.
9. Shandong daxue lishixi Zhongguo jinshi jiaoyanshi (Office for Teaching and Research on Modern Chinese History, Shandong University), Shandong Yihetuan diaocha ziliao xuanbian (Selected Survey Materials on the Boxers in Shandong, hereafter DCZLXB) (Jinan: Qilu shushe, 1980).
10. DCZLXB, pp. 13 and 18.
11. Ibid., p. 7.
12. Memorial of the Shandong Governor Tongxin, dated Jiaqing 18/10/7

(October 30, 1813), in the Grand Council Archives.

13. DCZLXB, p. 334.

14. Lao Naixuan, "Quanan zacun" (Miscellaneous Materials on the Boxers), and "Zhi Hu Shaojian shu" (Letter to Hu Shaojian), in Qian Bocan et al., eds., Yihetuan (The Boxers; hereafter YHT) (Beijing: Shenzhou guoguangshe, 1951), vol. 4, p. 459.

15. See Zhongguo shehui kexueyuan jindaishi yanjiusuo jindaishi ziliao bianjishe (Editorial Office for Modern History Materials of the Modern History Institute, Chinese Academy of Social Sciences), ed., Shandong yihetuan anjuan (Documents on the Boxers in Shandong; hereafter YHTAJ) (Jinan: Qilu shushe, 1980), p. 309.

16. Lao Naixuan, "Yihetuan jiaomen yuanliu kao" (On the Origin of the Boxer Sectarians), YHT 4:438ff.

17. DCZLXB, pp. 335-36.

18. Ibid., pp. 85-86.

19. Qing Gaozong shilu (Veritable Records of Qianlong Emperor [r. 1736-1796]), juan 1045, p. 35.

20. Yinxian zhi (Gazetteer of Yin County), compiled in the Xianfeng period (1851-1862), juan 29.

21. Jiang Kai, "Pingyuan quanfei jishi" (Record of the Boxer Incident at Pingyuan), YHT 1:354.

22. Confession of Fang Rongsheng in the Grand Council Archives.

23. "Qiuxin jiaofei" (Sectarian Outlaws at Qiuxin), Shandong junxing jilue (Account of the Military Operations in Shandong), juan 12.

24. Lin Xuezhen, ed., "Zhidong jiaofei diancun" (Collected Telegrams concerning the Suppression of Bandits in Zhili and Shandong), in Yihetuan yundong shiliao congbian (Collected Materials on the Boxer Movement; hereafter SLCB) (Beijing: Zhonghua shuju, 1964), part II, p. 70.

25. See Lao Naixuan's "quanan zacun" and "Zhi Hu Shaojian shu," YHT 1:459ff.

26. Jiang Kai, "Pingyuan jiaofei jishi" YHT 1:356.

27. DCZLXB, p. 139.

28. Ibid., pp. 213-14.

29. Li Di, Quanhuoji (Record of the Boxer Catastrophe), p. 346.

30. DCZLXB, pp. 150-51.

31. Ibid., pp. 192-94.

32. Memorial of the Shandong Governor Mingxing, dated Qianlong 51/8/9 (September 30, 1786), in the Grand Council Archives.

33. Memorial of the Shandong Financial Commissioner Zhu Xijue, dated Jiaqing 18 (1813), in the Grand Council Archives.

34. Yang Shurong, "Guangxu Genzi Wode jinzhongzhaotuan de zuzhi jingguo" (How My Golden Bell Unit Was Organized in 1900), an unpublished item in the Boxer Research Group's collection.

35. YHTAJ 1:365-66.

36. Ibid. 1:455.

37. DCZLXB, pp. 260-61.

38. YHTAJ, editor's introduction.

39. Li Di, Quanhuoji.

40. Chiping xianzhi (Gazetteer of Chiping County), revised 1935, juan 11.

41. YHTAJ 2:809; also see 1:435 and 925.

42. Weixianzhi (Wei County Gazetteer), revised 1923.

43. Sawara Tokusuke, "Quanshi zaji" (Notes on the Boxer Affair), in YHT

1:244.

44. Wanguo gongpao (Intelligencer of All Nations), September 1900.

45. Memorial of Zhili Governor-General Naerjinger, dated Xianfeng 1/11/26 (November 16, 1851), in the Grand Council Archives.

46. Memorial of Yinghe, dated Jiaqing 24/3/22 (April 18, 1819), in the Grand Council Archives.

47. Memorial of the Shandong Governor Tongxing, dated Jiaqing 19/3/23 (May 2, 1815), in the Grand Council Archives.

48. Memorial of Acting Zhili Governor-General Qishan, dated Daoguang 18/12/14 (January 28, 1839), in the Grand Council Archives.

49. Memorial of Salt Censor Wenli, dated Guangxu 26/5/15 (June 11, 1900), in the Grand Council Archives.

50. Memorial of Peng Jiaping, dated Qianlong 22/4/6 (May 23, 1757), in the Grand Council Archives.

51. Memorial of Shu Hechen, dated Qianglong 39/10/19 (November 22, 1774), in the Grand Council Archives.

52. Memorial of Acting Shaanxi Governor Yinwu, dated Qianglong 39/10/20 (November 23, 1774), in the Grand Council Archives.

53. Qing Gaozong shilu, juan 1072, p. 29.

54. Memorial of Shandong Governor Tongxing, dated Jiaqing 18/11/13 (December 22, 1806), in the Grand Council Archives.

55. Memorial of Shanxi Governor Zhu Xun, dated Jiaqing 20/8/22 (September 24, 1815), in the Grand Council Archives.

56. Memorial of Zhili Governor-General Nayancheng, dated Jiaqing 20/12/19 (January 17, 1816), in the Grand Council Archives.

57. Memorial of the Censor Qingpu, dated Jiaqing 19/10/14 (November 25, 1814), in the Grand Council Archives.

58. Ibid.

59. Memorial of Zhili Governor-General Nayancheng, dated Jiaqing 18/12/14 (January 5, 1814), in the Grand Council Archives.

60. Dai Xuanzhi, Yihetuan yanjiu (Researches on the Boxers) (Taibei: Wenhai chubanshe, 1964).

61. Fang Shiming "Yihe (quan) tuan yu bailianjiao shih liangge 'shitong chouhou' de zuzhi" (The Boxers and the White Lotus Were 'Deadly Enemies'), Shehui kexue jikan (Journal of Social Science) 180.4.

62. Memorial of the Shandong Governor Tongxing, dated Jiaoqing 18/9/11 (October 4, 1813) in the Grand Council Archives.

63. Ibid.

64. "Zixian Yanghu Wu Kai zhaomu guanding she" (Recruitment of Official Soldiers Issued by the District Magistrate Wu Kai), Jinxiang xianzhilue (Gazetteer of Jinxiang County), revised in Xianfeng 10 (1860).

65. Ibid.

66. Fang Shiming, "Yihe (quan) tuan yu bailianjiao."

67. Confession of Cui Huan in the Grand Council Archives.

68. Confession of Fang Rongshen in the Grand Council Archives.

69. Memorial of Cao Zhenyong, dated Daoguang 20/5/9 (June 8, 1840) in the Grand Council Archives.

70. Jindaishi ziliao (Materials on Modern History) 1957.1, p. 13.

71. Dai Xuanzhi, Yihetuan yanjiu.

72. Memorial of Shandong Governor Jingebu, dated Daoguang 16/9/7 (October 15, 1836), in the Grand Council Archives.

73. Xiang Da, "Ming Qing zhijizhi baojuan wenxue yu bailianjiao" (Con-

nections between Sutra literature from the Ming-Qing Period and the White Lotus), in Tangdai Changan yu xiyu wenming (Xian in the Tang Dynasty and Its Connections with the Civilization of Central Asia) (Beijing: Sanlian shudian, 1957).

74. Memorial of the Shandong Governor Jingebu, dated Daoguang 16/9/7 (October 16, 1836), in the Grand Council Archives.

75. Guowenbao (National News), Guangxu 23/20/6 (October 31, 1897).

76. Jiang Kai, "Pingyuan quanfei jishi," YHT 1:356.

77. DCZLXB, p. 221; YHTAJ 1:390.

78. YHTAJ 1:140.

79. DCZLXB, pp. 333-34.

80. Jiang Kai, "Pingyuan quanfei jishi," YHT 1:359.

81. Zhang Zhiqing and Li Peng, "Ronglu he Yihetuan ji qita" (Ronglu, the Boxers and Other Matters), an unpublished item in the Boxer Research Group's collection.

82. Yuyun, "Qing Duanjunwang Zaiyi fengjue hou zaji genggai" (A Sketch of Prince Duan's Official Career), an unpublished item in the Boxer Research Group's collection.

83. Huang Zengyuan, "Yihetuan shishi" (Facts About the Boxers), SLCB, part 2, p. 125.

84. George Nye Steiger, China and the Occident: The Origin and Development of the Boxer Movement (New Haven: Yale, 1927).

85. Dai Xuanzhi, Yihetuan de yanjiu.

86. Guojia danganju Ming Qing danganguan (Office of Ming-Qing Archives of the National Archives), Yihetuan dangan shiliao (Archival Materials on the Boxers; hereafter DASL) (Beijing: Zhonghua shuju, 1959), vol. 1, p. 15.

87. Letter of Shandong Governor Zhang Rumei to the Zongli Yamen, dated Guangxu 24/4/29 (June 17, 1898), quoted in Dai Xuanzhi, Yihetuan de yanjiu.

88. YHTAJ 1:138.

89. Guchun caotang biji, in the collection of Institute of Modern History of the Chinese Academy of Social Sciences.

90. DCZLXB, p. 329.

91. Weixianzhi, revised 1923.

92. "Guo Dongchen de qinbi huiyi" (Autobiographical Notes of Guo Dongchen), dated June 23, 1957. See Special Issue on the Boxer Studies of Shandong daxue wenke lunwen jikan (Bulletin of the College of Letters and Science at Shandong University), 1980, no. 1.

93. Ibid.

94. Chine et Ceylon, 1898. A French Catholic missionary journal.

95. Memorial of Censor Wenxiang, dated Tongzhi 1/5/27 (June 23, 1862), in the Grand Council Archives.

96. Shouzhang xianzhi (Gazetteer of Shouzhang County), revised 1901, juan 9, "Wulue" (Military Affairs).

97. Lin Xuezhen, "Zhidong jiaofei diancun," SLCB, part 2, pp. 44, 47-48.

98. Huibao, no. 146, dated Guangxu 25/12/10 (January 10, 1900).

99. See Steiger, China and the Occident (1927), pp. 131-32.

100. YHTAJ 1:17.

101. Ibid., pp. 13, 15.

102. Memorial of Shandong Governor Yuxian, dated Guangxu 25/11/4 (December 6, 1899), DASL 1:39.

103. Huibao 146 and YHTAJ 1:18 and 14.

104. Order of the Grand Council to the Shandong Governor, dated Guangxu

25/10/18 (November 20, 1899), in DASL 1:36-37.
    105. Xinwenbao, Guangxu 26/3/20 (April 19, 1900).
    106. Qiao Xisheng, Quanfei jilue (A Sketch of the Boxers), second part, final juan.
    107. DCZLXB, p. 127.
    108. Ibid., p. 126.
    109. Ibid., p. 127.
    110. Ibid.
    111. DASL 1:118.
    112. Memorial of Acting Zhili Governor-General Qishan, dated Daoguang 18/12/14 (January 28, 1839), in the Grand Council Archives.
    113. DASL 1; 156.
    114. Ibid., p. 207.
    116. Frederick Engels, The Peasant Wars in Germany (New York: International Publishers, 1926), p. 52.

JIN CHONGJI

The Relationship Between the Boxers and the White Lotus Sect*

I want to discuss some ideas concerning the question of the relationship between the Harmony Boxers (Yihequan) and the
White Lotus sect. My general thesis is that there was some relationship between the two, but they were not the same. Simply
stated, the Harmony Boxers dealt with boxing while the White
Lotus sect was a religion. This reflects their basic differences.
Of course, there was some interaction between them, but their
complex connections should not make us mistakenly conclude
that they were the same.

Everyone knows that the White Lotus sect had its own religious faith and a collection of religious principles. It had its
own sacred texts and believed in both "Our Homeland of True
Emptiness" (*zhenkong jiaxiang*) and the "Eternal Mother"
wusheng laomu). Although the Harmony Boxers had tinges of
superstition, they did not become a religion. Religion is, of
course, a superstition, but not all superstitions are equivalent to
a religion. The Harmony Boxers' superstitions were linked to
their boxing practices. Their invitations to the spirits to occupy
their bodies were for the purpose of self-protection, as we can
see from the chant "Swords and spears will not penetrate
(*daoqiang buru*), and their secret rhymes, such as

Kowtowing toward the North opens the Cave gate.
From it summon the Iron Buddha upon you to wait.[1].

In the end, these chants had deteriorated to "Stop the guns and
the bullets cannot come."[2] They imagined that firearms would

---

*From Yihetuan yundong shi taolun wenji (Collected Articles on the History of
the Boxer Movement) (Jinan: Qilu shushe, 1982), pp. 27-33. Translated by James
Bollback.

not cause wounds because of the protection provided by the spirits who had possessed one's body.

Because the White Lotus sect was declared a "heterodox sect" (*xiedao*), its organization had to be relatively closed and it had many secret activities. The White Lotus had strict admission procedures and rituals; it held periodic meetings and required the payment of dues. It had a sect master (*jiaozhu*) and a hierarchical structure. Tao Chengzhang, in *Jiaohui yuanliu kao* (Research on the Origins of Religious Sects and Secret Societies), wrote: "The power of the sects is linked up, so even though they may be dispersed over several thousand *li*, they cannot be controlled . . . they function just as a body makes an arm work, or the arm makes the fingers work."[3]

The Harmony Boxers were not like that. In each locality they were a comparatively open association for the practice of boxing. They had a mass, popular nature. The purpose of the boxing, at first, was to build a good physique; to protect oneself and one's family. They opened public boxing grounds so that whoever wanted to join could come, and whenever anyone wanted to quit they could. Relationships were equal among the boxing participants. Elderly informants told us that "Those who learned Spirit Boxing (*shenquan*) were all the same. There wasn't any difference between them. There weren't any chiefs and followers; there were only those called 'Elders' (*dashixiong*). An Elder was one who had done well in the practice of boxing. Anyone who did well could be addressed as 'Elder.' " Every village had a boxing ground where they could compete among themselves, but they lacked any hierarchical relationships. This shows some of the differences between the two groups.

During field investigations at Chiping county in Shandong, the elderly people said, "The White Lotus Sect and the Spirit Boxers were not the same. The White Lotus could summon wind and rain, or could take a stool and turn it into a horse to ride." Others told us that "The White Lotus sect was something different. At Wuguantun there was also a White Lotus sect. . . . They wanted to rule the earth, to start a dynasty and select a queen. The Bald Girl (Tu Guinu) was chosen as the Queen. . . . Their revolt was put down within a day. That is why people say, 'When Wuguantun revolts, it is only a one-day affair.' "[4]

If we study the movement even more concretely, the Harmony Boxers can been seen to have undergone a process of maturation. By the time the movement reached its peak, people for many dissimilar reasons had taken up the Boxers' cause. The Boxers were different in their early period from what they were in their later period; they differed in one county from another. Even within the same region, their circumstances were not exactly the same.

We can describe at least three periods in the Boxers' development.

I

Everyone agrees that the earliest incidence of use of the name Harmony Boxers came in the struggle against the foreign churches known as the Liyuantun affair in the Guan county exclave. That branch was led by Zhao Sanduo. There are three notable characteristics of this branch. First, this branch of the Harmony Boxers originally was called "Plum Flower Boxers" (Meihuaquan). It had been in existence in the border regions between Zhili and Shandong for a long time, but in the past it had not been involved in any rebellious activity. When Zhao Sanduo wanted to lead the Plum Flower Boxers against the foreign churches, he encountered opposition from some of the local Plum Flower Boxing leaders. They said: "Don't start a great disturbance. From the time of our founders in the late Ming and early Qing there have been sixteen or seventeen generations of teachers. On the civil side (*wen*) we have burned incense and treated the sick; on the martial side (*wu*) we have practiced boxing to build up our physiques, but there has never been a rebellion. If you rebel and fail, we cannot go around in public." Zhao Sanduo is said to have replied: "I am riding on the back of a tiger and cannot get down. I don't dare stop, for the Catholic Church will never forgive me. I'll change the name of my branch of the Plum Flower Boxing to 'Harmony Boxing' (*yihequan*) and will separate myself from you."[5] Thus, Zhao Sanduo changed the name of his branch.

Second, this branch of the Harmony Boxers developed relatively early, but it only practiced boxing and did not burn incense or chant incantations, and it had few superstitious characteristics. Third, the slogan "Assist the Qing; Eliminate the Foreigners" *zhu Qing mie yang*) was first used by this branch. A veterinarian called Venerable Zhang, who was attached to the Qing troops at Liyuantun, advised Zhao Sanduo to use this slogan. "Then," he said, "the government authorities will not have to investigate who is to blame." Zhao Sanduo accepted his suggestion and wrote on his banners, "Assist the Qing; Eliminate the Foreigners; Slaughter the Catholics" (*shajin tianzhujiao*). After the Liyuantun incident, the anti-Christian reputation of the Harmony Boxers began to spread. Still, we cannot locate any relationship between the Harmony Boxers there and the White Lotus sect.

II

The second period was when the Harmony Boxer name was used in Chiping, Gaotang, and Pingyuan counties. They began to attack church buildings and to oppose Christianity. The most famous person among them was Zhu Hongdeng (literally, "Red Lantern" Zhu). Not even Jiang Kai, who was the county magistrate in Pingyuan at the time, was clear about where this branch of the Harmony Boxers came from. In his book *Pingyuan quanfei jishi* (A Record of the Boxers in Pingyuan), he wrote, "I have heard that all of En county is rampant with these Boxers. Some say they came from the Shibacun exclave of Guan county, and others say they came from Dongchang county, Caozhou prefecture, but no one is certain."[6] At this same time the children of Chiping county sang this ditty:

First study the Plum Flower Boxing,
Then study the Golden Bell.
Kill the foreign devils and
Destroy the Catholic Church.

The Armor of the Golden Bell (Jinzhongzhao) was the same as the Big Sword Society (Dadaohui), whereas Dongchang was the place in Caozhou prefecture where the Big Sword Society originated. The so-called Eighteen Militia Units of Shibacun were simply the branch of Plum Flower Boxing led by Zhao Sanduo. These two pieces of information reflect the origins of the Chiping and Pingyuan area Harmony Boxers from a combination of the Plum Flower Boxers and the Big Sword Society.

There are also three points concerning the Caozhou Big Sword Society that warrant attention. First, at the very beginning, the Big Swords had the character of a united village organization (*lianzhuanghui*) in that they were led by landlords and the members were all owner-peasants. That is to say, most of the members were people who owned land and had their own households. Yet some tenants also joined, primarily to protect their landlord's property. In this region, which was a border area of the four provinces of Shandong, Jiangsu, Henan, and Zhili, the local bandits were particularly numerous. The Big Swords were organized primarily to protect the property of the landlord class. The *Shandong shibao* (Shandong Times) carried an article entitled "Stopping the Spread of the Big Sword Society," which stated: "The people suffer at the hands of bandits and thieves and have no way to deal with the problem. They heard that the Big Sword Society could protect them against sword and shot, and so enable them to protect themselves and their families. Thus they engaged in a frantic competition to study, regardless of cost. The rich families who follow its teaching number in the thousands."

Liu Shiduan, founder of the Big Swords in Caozhou, came from a family with more than one hundred *mu* or about seven hectares of land and was the wealthiest person in the locality. It was he who established the Big Swords in Caozhou. Second, the most important practice of the Caozhou Big Swords was their swordsmanship. They had a teacher, Zhao Jinhuan, who was invited to come especially to teach swordsmanship. It was said that this man had escaped when the White Lotus sect was broken up, and that he was still a member of the White Lotus. He was invited to come by Liu Shiduan, and in the process of instructing in swordsmanship, he also introduced the practice of "calling spirits to possess a person's body" (*jingshen futi*). This gave the Big Swords' practices a decidedly superstitious coloration. Perhaps that is what Zhao Jinhuan, as a disciple of

the White Lotus, contributed to the Big Sword Society. He was
a teacher who had been invited only to teach swordsmanship
and not to propagate his religion, and in the end the Big
Swords did not become a part of the White Lotus sect because
of him. Finally, the Big Sword Society had begun for the purpose of opposing local bandits; later it expanded to include attacks on churches. In the beginning the attacks were the result
of internal contradictions within the landlord class.[7]

The first large-scale activity of the Big Swords was when
Liu Shiduan and others responded to the invitation of a rich
man from Pangjia linzhuang in Dangshan county, Anhui province. Pang Sanjie, whose family had more than three hundred
mu or about twenty hectares of land, asked for help in his fight
to take some marshland from the Dongtuan Christian church,
which wanted to reclaim the same piece of land. As a consequence, Liu Shituan and others were killed by Qing dynasty
officials. After this the anti-Christian struggle of the Big Sword
Society began to develop even more rapidly.

Zhu Hongdeng's branch of the Harmony Boxers spread
northward, from Changqing to Chiping, and then to Gaotang
and Pingyuan counties. In the beginning it was called "Spirit
Boxing," but later it took the name Harmony Boxers and used
the slogan "Promote the Qing and Eliminate the Foreigners"
(*xing Qing mei yang*). Some of the common people of the
northern region called them "the Big Sword Society." From this
we can see that the name of this branch of the Harmony
Boxers and their slogans derive from Guan county exclave's
Plum Flower Boxers. However, Zhu Hongdeng's boxing practices and his superstitious techniques, including the demonstration by a boxing master, derived primarily from the Big
Sword Society in Caozhou prefecture.

Zhu Hongdeng's branch differed from the Caozhou Big
Swords in another respect. His followers--unlike the Caozhou
Big Swords who were composed of men directly controlled by
the large landlords--were mostly poor peasants, vagabonds, and
handicraft workers, joined by a few middle or small landlords.
Although Zhu Hongdeng's group had heavy superstitious overtones, there were fewer religious elements of White Lotus derivation than in the Caozhou Big Sword Society. In addition, Zhu

Hongdeng's group was much more loosely organized and not as strict as those in Caozhou.

The battle of Senluo Temple in Pingyuan is the most famous of Zhu Hongdeng's activities. In later days the real circumstances of this battle were greatly exaggerated. Zhu Hongdeng did not lead a large group of mounted men from Chiping to Pingyuan; he brought only ten to twenty men. The rest were local people from Pingyuan who temporarily had been recruited from boxing adherents (*quanmin*) of that region. Zhu gathered more than one thousand men and attacked the church at Liuwangzhuang. Passing through the settlement at Senluo Temple, they encountered General Yuan Shituan's government troops. They were preparing to greet the troops on friendly terms, but Yuan's soldiers opened fire. The enraged Boxers then rushed the army. As a consequence more than ten government soldiers and seventy to eighty Boxers were killed. Zhu Hongdeng's men continued to fight with the soldiers, but the local people who had joined his forces simply fled. Zhu Hongdeng had no choice other than to return to Chiping with his original small group.

III

The influence of this incident, however, was great. Henceforth the reputation of the Harmony Boxers for attacks on Christians spread near and far. Even though it was not long before Zhu Hongdeng was executed by the Qing government, the Harmony Boxers' popularity grew broader and they spread into Zhili. This battle at Senluo Temple may be considered to have inaugurated the third stage in the movement. This stage was not in the style of the Taiping Heavenly Kingdom's military expedition. There was no such thing as an army of Shandong Boxers moving northward along two main routes. Rather, it arose from among the Zhili people who had been oppressed by foreigners, or who had been cheated by the Christian church. Aroused, they flocked to the banner of the Harmony Boxers and started the struggle against the foreign churches.

At this point the struggle became more complex. In Zhili there were some White Lotus sect members who "had no connection" with the Harmony Boxers but began using their name for their own groups. According to material from Guo Dong-chen, when Zhao Sanduo held a meeting in the Temple of the Great Buddha (Dafoshi) in Zhending, someone named Li from Hejian prefecture said:

At present in the counties of Jinghai, Qing, and Leguang [in Zhili] there is a Red Gate (Hongmen) organization that sec-retly practices the eating of charms and the chanting of in-cantations. They are like the Iron Shirts (Tiepushan). . . their boxing is not done publicly and they follow the teachings of the White Lotus sect. At this time the common people have a great deal of trust in them. Their goals are the same as ours. They have no connection with us, but it would not be bad if we moved a bit closer to them. . . . We Harmony Boxers are open. We hold meetings and practice boxing any-where. If they linked up with us, and could also openly and publicly practice boxing and hold meetings, then certainly they would want to unite with us.[8]

Indeed, after the Harmony Boxers spread into Zhili, the influ-ence of the White Lotus sect became much deeper than in the Shandong period because many people joined the Harmony Boxers who had formerly been White Lotus sect believers. But this was not true in all instances. In Zhili there were also those White Lotus sectarians who opposed the Boxers. In investigative materials from Anci and Wuqing counties in Hubei, Zhang Shi-jie relates that in Luofa there was a rich landlord who said that "in the name of the White Lotus sect, he could bring to life figures made out of paper, both men and horses. He could, furthermore, turn soybeans into soldiers. He roped in members of the White Lotus to destroy the Boxers, but they were easily eliminated by the Boxers." Material is often cited relating that in Beijing the Boxers had proclaimed that "the White Lotus sect has entered into an agreement with the Catholic Church and will revolt on the fifteenth day of the eighth month." This is said to have caused the death of a large number of people whom the Boxers said were White Lotus followers.

From this simple investigation we can see that the Harmony

Boxers had some relationship with the White Lotus, but they were not the same. Then why has the theory that the Boxers originated from the White Lotus sect or that the two were the same had such a great influence and survived for eighty years? I think that there are some political reasons, as well as some grounds created by the complex nature of the facts.

First, let us mention the political reasons. At the time, Lao Naixuan, the county magistrate at Wuzhao in Zhili, wrote *Yihetuan jiaomen yuanliu kao* (Investigations of the Origins of the Boxer Sectarians), and he wanted to emphasize the idea that the Harmony Boxers had derived from the White Lotus sect, because the Qing government considered the White Lotus to be a heterodox religious sect. His objective was to charge the Harmony Boxers with being a heterodox sect in order to redouble the official suppression. In the same way, Jiang Kai wrote in *Pingyuan quanfei jishi* that when General Ma Jinxu of the Qing forces captured Zhu Hongdeng "ten private letters" were uncovered that mentioned "an attack on Beijing on the eighth day of the fourth month, next year." This statement is often quoted even today, but I am almost certain that this was fabricated by Jiang Kai. Zhu Hongdeng's power was very limited at this time; he did attack churches in his home area, but only when the imperial government began to suppress him did he do battle with government troops. He could not have hoped to mount "an attack on Beijing on the eighth day of the fourth month, next year."

The volume *Shandong Yihetuan anquan* (Files on the Boxers in Shandong) contains contemporary reports from Ma Jinxu, as well as the report of the Jinan prefectural magistrate who was in charge of interrogating Zhu Hongdeng.[9] Neither of these mentions the "private letters." It seems to me that these letters were created by Jiang Kai in order to add to the charges against Zhu Hongdeng. After Liberation, many of our comrades made the mistake of citing this material, but for the opposite purpose. They wanted to praise the peasants' struggle. They felt this material could explain the long tradition of the Harmony Boxers' resistance to the Qing government.

On the other hand, the complex nature of the connections is important in itself. Some branches of the White Lotus sect such as the Heavenly Doctrine Society (Tianlihui) also practiced boxing. Furthermore, there were also some White Lotus followers who joined the Harmony Boxers throughout their history. These mutual relationships mean the two influenced one another, and so it becomes easy to confuse them. In addition, the Qing government had prohibitions against both religious sects and boxing groups. Both fell into the category of illegal practices, but careful investigation shows there were still important differences between the two. The government strictly forbade religion, but prohibition against boxing was not as severe, so that in many places the practice of boxing could be rather public. Thus, the fact that in the official documents the Harmony Boxers were mixed up with the White Lotus is not strange, because the government itself did not always clearly differentiate between the various kinds of organizations in the lower social strata. Similarly, in the beginning of the Taiping Heavenly Kingdom the documents of the Qing officials sometimes confused the God-Worshipping Society (Baishangdihui) with the Heaven and Earth Society (Tiandihui). These certainly cannot be taken as proof that the God-Worshippers sprang from the Heaven and Earth Society or that they were a branch of that society.

What value does the clarification of this issue about the relationship between the White Lotus and the Harmony Boxers have? In the first place, the study of history should begin with the facts. After clarifying them, interpretation can be made. In the second place, as a result of this procedure, we can see one special characteristic of the Boxer movement: it had a broad mass nature.

One of the fundamental reasons why this movement could manifest such great power was that it sought to unite with broad masses and go beyond the secret societies. It was based directly within the great peasant masses. In addition, it reflects the fact that movements or organization among the lower ranks of society did not really contain much political coloration. Nonetheless, when the people suffer the bullying ruthlessness of foreign aggression, such organization can also be transformed and become the means of mass resistance and struggle, which may develop great strength. Obviously, the Harmony

Boxers developed in the face of imperialist aggression, yet they also had a backward and ignorant side. They were sometimes indiscriminately antiforeign and did not understand the nature of the Qing state. Nor did they offer a program of scientific struggle. This was unavoidable. Today, when we recall the original features of the Harmony Boxers, we do not need to avoid recalling their faults.

Editor's Notes

Jin Zhongji did not provide any references in his essay. These have been added to help understand the translation or to help locate some of his evidence.

1. Qishou beifang tongkai/tong zhong qing tiefo lai.
2. Zhizhu qiang pao bunenglai.
3. This piece by Tao Chengzhang [d. 1911] is reprinted in Chai Dengeng et al., eds., Xinhai geming (The Revolution of 1911) (Shanghai: Renmin chubanshe, 1957), 3:99-111.
4. Shandong daxue lishixi, Zhongguo jindaishi jiaoyanshi (Chinese Modern History Teaching Materials Section, Department of History, Sandong University), Shandong Yihetuan diaocha ziliao xianbian (Selected Survey Materials on the Boxers in Shandong; hereafter DCZLXB) (Jinan: Qilu shushe, 1980), pp. 128ff.
5. DCZLXB, pp. 264-68.
6. Qian Bocan et al., eds. Yihetuan (The Boxers) (Beijing: Shenzhou guoguangshe, 1951), 1:327-32.
7. DCZLXB, pp. 6-22.
8. Guo Dongchen was a local cultural worker who wrote his recollections of the Boxers in 1960 when he was about eighty. See "Yihetuan zhi yuanqi" (The Genesis of the Boxers), DCZLXB, pp. 327-32.
9. (Jinan, Qilu chubanshe, 1980).

LI JIKUI

How to View the Boxers' Religious Superstitions*

The Boxer Uprising was a spontaneous mass movement against the imperialist Powers who were attempting to partition China. As an armed struggle, it exhibited fanatic religious superstition and indiscriminate antiforeignism. Even today, historians still differ in their judgment about these two qualities of the Boxers. We need to analyze the Boxers' antiforeignism and their religious superstitions because these questions have a bearing on the character of the whole movement. The present article, however, will be confined to the issue of religious superstition.

I

The Boxers were an anti-imperialist, patriotic organization that emerged at a critical juncture of the Chinese nation. They were not a genuine religious movement, for religious superstition was not central to their character.

In nineteenth-century China, the Christian churches played the role of herald to invasion by the Powers. The missionaries and their lawless native converts took advantage everywhere of their privilege in order to do evil. An official in Shandong reported: "Recently the Christians have become increasingly arrogant. . . . They are domineering in the villages, oppress the good people, and even force officials to do their bidding. They are such bullies that neither the officials nor the common people dare to resist them. . . . The accumulated grievances of the villagers often lead to serious conflicts. Yet the missionaries

*From Yihetuan yundong shi taolun wenji (Collected Articles on the History of the Boxer Movement) (Jinan: Qilu shushe, 1982), pp. 220-34. Translated by Zhang Xinwei.

will only listen to one side of the story and do not inquire into the true causes of the disputes. In minor cases they demand damages; in major ones, they apply coercion until we yield to their desires and thereby extort many advantages."[1] The people, unable to stand their oppression, gradually shifted their resentment away from the Qing government and toward the Christian churches. From 1890 onward, religious cases appeared successively in the prefectures of Caozhou, Dongchang, and Jinan, where churches were concentrated. By 1900, the Boxer movement had spread over China like wildfire.

The Boxers had evolved from various boxing societies, mainly the Big Sword Society (Dadaohui) in southwestern Shandong and the Spirit Boxers (Shenquan) and Red Boxers (Hongquan) in northwestern Shandong. In terms of organization, ideology, and activities, these Boxers had some association with the secret White Lotus sect, which had been active in Shandong and Zhili for a long time. They were "similar in behavior to the religious bandits of the Jiaqing reign (1796-1820)". But there was a difference in that the Boxers were not an antidynastic organization, but rather one intent on national salvation at a time when the nation's very existence seemed in peril.

According to contemporary records, "The Big Sword Society has existed for a long time. Popular accounts have it that its magic can ward off bullets and shells. Its charms were practiced by many people in Zhili, Shandong, Henan, and Jiangsu. In recent years, frequently after being bullied by native Christians, more people have joined in the drills [of boxing groups] to protect their lives and property. Another society, called the Red Boxers, also practices the art of boxing for the same purposes. This is the background of these two groups."[2]

Originally there was a difference in purpose and activities among these boxing societies. The Big Swords, also known as the Armor of the Golden Bell (Jinzhongzhao), were a superstitious organization. Members "relied on magic, chanted incantations, and swallowed charms to make themselves impervious to swords and bullets. This magic could be learned overnight. With its protection, one could not be hurt by cudgels, swords, bullets, or even cannonballs, because the believer had *gongfu* all around his body. . . . On hearing the news, the common people, harassed by armed criminals, rushed to learn this magic. Their numbers quickly mounted into the millions."[3]

In the beginning the Spirit Boxers were formed for physical fitness and defensive purposes. They primarily practiced boxing and had their own simple statement of ideology: "Be filial, fraternal, and loyal. Respect the old, love the young, and be kind to your neighbors. Abstain from wine, lust, greed, and bad temper." They were also enjoined from harassing common people.[4] "People in the vicinity of Chiping county, outraged by the oppression from native Christians, started to practice boxing. They said the gods had taught them this boxing for the purpose of resisting Christianity."[5] Apparently, these mass organizations were originally formed for self-preservation and were armed only with simple weapons such as swords, spears, and cudgels. These were inadequate to resist the Christians' power. Something was needed to make up for their inferiority to the foreigners.

Becuase of the lack of material weapons, the Chinese had recourse only to spiritual defenses. Thus, the boxing societies became more and more superstitious as time passed. In the *Chiping xianzhi* (Chiping County Gazetteer) it is recorded, "After the Sino-Japanese War, the people became more angry, for they felt that the foreigners had won because of their guns and cannons, and there must be some way to counter them. Hence the tales of Armor of the Golden Bell' supernatural powers to protect those within its compass. At first the Golden Bell ritual was practiced secretly in one or two places. Then it spread to all the villages where boxing grounds had been set up."[6] The sudden rise of such religious superstitions can be explained by the saying: "Patriotic and yet not knowing what to do, the illiterate people and youths looked to the gods for help."[7]

Because they had a common aim, it was quite natural for the boxing societies to cooperate with one another and, in some cases, to merge. It became increasingly difficult to distinguish between them, and so they were given the general designation of Boxers. "In May 1900, Zhu Hondeng, leader of bandits from Changqing, recruited people from Pingyuan and En counties and established a society known as the Harmony Boxers (Yihe-quan). This society previously had used names such as Red Lanterns, Armor of the Golden Bell, Iron Jackets, and Big Swords. Suppressed and pursued by government troops, they fled to Pingyuan and En counties and changed their name first into 'Yihequan' or what is now called 'Yihetuan.' "[8]

The basic unit of Boxer organization was called an "altar" (*tan*), and these units were not subordinate to another, for there was no supreme command or overall leader. Their religious beliefs were polytheistic. They lacked any source of authority for their beliefs and, in fact, did not believe that one religious doctrine should exclude others. The deities they worshipped were a heterogeneous lot. Some were Buddhist, such as Sakya-muni or the Goddess of Mercy; others were Daoist, including the Jade Emperor, the Ancient Ancestor Hong Jun, or Zhang, the Heavenly Teacher. Some of these were figures traditionally respected by Confucians, such as Jiang Ziya, Zhuge Liang, Duke Guan, and Liu Bowen. Others were well-known characters from plays or supermen from folklore.

Some Boxer leaders took the names of those deities and his-torical figures as their assumed names. The Boxers spread tales about supernatural powers, wrote charms, swallowed elixirs, and practiced magic writing and performed rituals to invoke the spirits to possess them. These religious superstitions were used to mobilize the masses and encourage them to fight the foreigners. Contemporary records show,

> By custom those who join these groups only practice boxing and other arts of self-defense, but actually they have learned incantations that are supposed to enable spirits to possess these followers. Once possessed, the followers wield spears and cudgels like madmen.
>
> When they plan to attack a village, they will send out notices in advance. After their supporters have assembled on the scheduled day, each person will swallow a charm and chant incantations. Then, as they dance in a group, incense is burned to invoke the spirits. The leader then issues orders. To mislead his followers, he will pretend that the spirit is speaking through him.

In battle [the leader] holds a yellow banner or wears a yellow robe with the insignia of a spirit on the back. The ranks are armed with spears, swords, birdguns, and native-made cannons. They kowtow toward the southeast and mutter spells before they go into battle, believing that they have been made invincible.[9]

Many of [the Boxers] are boys of fourteen and fifteen, who were induced to join at the start and then forced to continue under duress. When the society calls a great gathering, its members travel hundreds of miles to attend.[10]

In the capital, many of those who practice Boxing are mere children, but also officials, peasants, workers, merchants, and tradesmen all want to learn it. Having studied the proper incantations from a teacher, one can be possessed by a certain spirit or deity. Then one becomes proficient in the martial arts, agile in one's movements, and skillful with sword and lance. The neighborhoods with boxing grounds are too many to count.[11]

In North and Northeast China, the Boxers grew rapidly in numbers. "In northern Zhili around Baoding, Zunhua, and Jingzhou, the rumor ciruclated that a spirit had descendend to the world in the form of a teacher. This teacher took only children as his disciples and taught them spells, boxing, and swordplay. When they were proficient enough, they would become invulnerable to swords and guns. Thus the Boxers became stronger with each passing day. Even adults started to believe and to join them, until every village had its own altar. Thus it was that the banner of the Boxers first came to be raised. Later, still more people were recruited until Boxers could be found almost everywhere in the three northern provinces."[12]

The movement became increasingly hallucinatory with magic arts, incantations, the reincarnation of Liu Bowen, the prophesies of a wandering Daoist priest, and tales of a Great Teacher and the Red Lantern with their supernatural powers. Exaggerated descriptions from some accounts made the movement even more bizarre. This led to the Boxers being condemned by educated men as "a stupid and ridiculous razzia" or a "general mobilization of reactionary forces" whose only aim was destruction and who lacked any constructive goals. This was a distorted view, for we know that under feudalism it is not only in China that true peasant movements may arise under the banner of religion.

In China such struggles involving religious fanaticism seemed to have reached their fullest development. Examples can be found in the Taiping Rebellion, where the insurgents fanatically believed in the religion of the God Worshippers, while their leaders claimed to be the spokesmen for the Heavenly Father, the Heavenly Elder Brother, and the Imperial God, who had descended into the world. To people with a knowledge of science, these tales of spirits and ghosts seem utterly preposterous, but in a country rife with illiterates, such tales form an important element in social life. Especially when the people can find nothing in real life upon which they can rely to overcome their difficulties, they tend to indulge themselves in fantasies, hoping for revelations or help from spirits possessing supernatural powers.

This was true in the long-stagnant feudal society of China, especially in the countryside of North China where economic backwardness and the cruel exploitation by the ruling class inevitably produced backwardness and parochialism among the peasants and the small-producer class. Thus, poverty and ignorance contributed to the Boxers' religious superstitions.

In Shandong and Zhili, where the movement originated, there was a long tradition of forming secret societies. Rebels had always used religion to organize people in their struggles. Undoubtedly, the Boxers appropriated this tradition. "The so-called Boxers are but peasants."[13] "At first, those who joined a Boxer unit were only ignorant peasants, oppressed and bullied in ordinary times and now without anywhere to turn with their grievances. When they heard the tales about the Boxers, they believed they might rely on new supernatural powers to vent their accumulated anger. So, people left their farm work to join the Boxers in the towns."[14]

In the Boxer ranks were people from the lower classes of society such as urban tradesmen, peddlars, soldiers, and vagrants, all of whom suffered from both domestic oppression and foreign aggression. These people were unemployed and penniless, ignorant and yet wedded to monarchical notions. Without the support of a progressive class, these people could look only to their predecessors for inspiration. Naturally their movement incorporated strong elements of religious superstition. Yet since we consider a struggle for national salvation against foreign aggression to be a just cause, the Boxers should not be condemned simply because of their backward ideology or the primitive form of their struggle.

II

The religious superstition of the Boxers became a means for furthering their struggle. It also reflected the strong antiimperialist will of the peasants of North China.

Once the Boxers appeared, various elements of the ruling class referred to them as "the outlaw Boxing groups" (*quanfei*) or sometimes "the righteous people" (*yimin*). From these differences developed the controversy in the Qing government about whether to adopt a policy of suppression and pacification toward the Boxers. These differences produced different policies as the movement developed. Of course, the Qing ruling clique's ultimate aim remained suppression of the movement.

Historical records indicate that until the movement reached its high tide, its ranks were relatively pure. The Boxers had a definite purpose and good discipline. An official of the Qing government admitted, "although these boxing societies have a large membership, there are no reports of them harrassing people or attacking government offices. Occasionally they have troubles with the Christians . . . but these are not occasions that require bringing in our troops."[15]

In Zhili, "it was learned that at Shuangliushu village in Zhuozhou, the Boxers fled after wounding two native Christians, and before taking anything from the houses. Their behavior emulated that of the knight errants in ancient times. This shows, then, that these men were not diehard bandits. Their destruction of the railway and telegraph lines was due to their mistaken understanding that these were built by foreigners. As a matter of fact, they do not know that these were built by the Qing state."[16]

Even the Guanxu emperor's "Edict of My Crimes" admitted: "After the Boxers in Laizhou and Zhuozhou had burned down churches and destroyed the railway, the regular army was dispatched to suppress them. Our troops, however, failed to act in a disciplined manner and massacred loyal people. The Boxers did not make trouble in their local villages and declared that they only hated the Christians. Consequently, the common people feared the soldiers, but came to praise the Boxers."[17]

Guan He wrote about the situation in Tianjin during these times: "[The Boxers] neither drank tea nor ate meat. Their daily food, called "victory rolls" (*deshengbing*) was supplied by shopkeepers."[18] The anonymous "Record of a Month in Tianjin," in its detailed account of the Boxers' activities, records, "In each altar there were many tablets with deities' names on them. . . . A candidate for inititiation would first kowtow to the tablets and vow never to recant. Then the teacher would instruct him: 'Abstain from greed and lust. Obey your parents and abide by the law. Exterminate the foreigners and kill corrupt officials. On the streets tend to business and do not dawdle; bow when you meet a comrade.' " The same book relates that when an altar was first set up, the candidates for membership had to have a reliable guarantor. The candidate must kneel before the altar and make his vows. Individuals unwilling to join could not be forced to do so.[19]

In Beijing, some Boxers received their supplies from the important protectors. Nobles, officials, businessmen, and wealthy families would make donations to the altar unit for their own protection. Some Boxers lived by their own means and had "only millet porridge or corn buns for their meals." Boxers "all stayed with their units which occupied the quarters of unused temples or inns."[20] It was said that they "sought neither fame nor fortune and acted without regard for their personal safety. They went into battle fearlessly, united by a common purpose." On the march the Boxers' "ranks were kept in better order than those of regular troops."[21]

Discipline was maintained by the Boxers' devotion to asceticism. For a considerable time, they maintained strict discipline, showed high morale, and acted resolutely. Before a number of altars undertook a joint operation, "notices would be sent out and soon thousands of people would assemble, well-armed with swords and spears just as the regular troops." Boxers from a radius of one hundred kilometers or so would assemble. At the sight of such a spectacle, a Qing official said in dismay, "[The Boxers'] forces number in the thousands and are well-armed. They threaten to burn and kill at the slightest provocation, while the officials just look on, not daring to do anything about them. Proper order does not exist."[22] These facts prove that the Boxers, in their early days, were a well-organized, combat-worthy force rather than a mob who beat, smashed, looted, burned, and killed at will.

The Boxers put up placards, made inscriptions, used slogans, and spread ballads to advance their causes. In just a few months, North China was thrown into an uproar. The inertia and silence of the past were suddenly swept away. Their propaganda materials were full of militant antiforeignism. Some read: "The spirits help the Fists, the Boxers, because the devils disturb China." "The Heavens are without rain and the earth is parched, because the churches block Heaven." "We are neither heretics nor White Lotus; we sing incantations and follow the truth." "When all the arts of war are fully learned, it will not be difficult to defeat the foreigners." "Tear up the rails and pull down the telegraph poles. Then destroy the steamships. France will tremble with fear; England, America, Germany, and Russia all will flee."[23]

The Boxers emphasized "the most hateful of all was the 1895 Peace Treaty whose consequences still can be felt today. The cession of territory and the payment of indemnities have been calamities for the nation and the people."[24]

The Boxers even set a "day of doom" for the foreign aggressors, declaring that the enemy would be completely annihilated within three months. Some even claimed this could be accomplished and peace restored by November 1900. Such materials need not have been produced by peasants, because among the Boxers were also people from well-to-do families and intellectuals. Their calls for millions of "spirit soldiers" to descend to wipe out the foreigners and create a new millennium undoubtedly reflected an intense anti-imperialist patriotism and a clear expression of the accumulated rancour of the millions who had suffered from foreign oppression.

Their patriotism made them utterly fearless. They shared a bitter hatred for their enemies and believed themselves to be invulnerable to bullets, "and so vied for a place in the van of each action. When those in front were mowed down like grass by the guns of the Allied troops, those in the rear would follow on undaunted in the face of death, even though they too would be shot and killed." Their bravery was incomprehensible to the "Westerners who pitied their ignorance."[25]

Some people interpret the militancy of the Boxers as an expression of their ignorance and feel that only because of superstition did the Boxers rely on spirits and incantations in battles. This is untrue, for the Boxers were motivated by an utterly fearless and selfless patriotism, which is a distinguishing characteristic of the modern Chinese peasantry in its struggle against aggression. Intense anti-imperialist fervor strongly shaped their religious superstition.

Because they were foolish enough to oppose the foreigners' firearms with hand-held weapons, the Boxers certainly lacked wisdom. In battle as their charms and incantations failed and their dead mounted around them, anyone in his right mind would comprehend that no spirits could protect him and would realize that magic arts were worthless. Did the Boxers really believe that the foreigners could be defeated with charms and incantations and that their own persons were impervious to guns? Of course not. The Boxers' chants in battle were similar to taking a vow to destroy the enemy or die in the attempt. They served as a means to boost morale, arouse hatred for the enemy, and heighten confidence in victory. In every age warfare requires special spiritual strength. In certain situations this spiritual commitment can be a decisive factor. It was such a commitment that led the Boxers to attack shouting "Kill all the foreign devils! Restore peace throughout the great Qing empire."

Furthermore, the Boxers did not simply rely on supernatural forces to overcome the foreigners. They used violence and were armed "with spears, swords, birdguns, native-made cannons, and even foreign guns. It was not religious superstition, but the general desire to oppose aggression that drew tens of thousands into the Boxers' ranks. We cannot even discuss the positive, anti-imperialist character of the Boxer movement unless we recognize that the Boxers are not to be ridiculed. Scholar-officials of the old order said of the Boxers: "Bare-handed and muttering spells, they went into battle only to be massacred by gunfire,"[26]

Even foreign observers such as Robert Hart, the inspector-general of the Imperial Maritime Customs Service, wrote: "[The Chinese nation] has slept long . . . but it is awake at last, and its every member is tingling with Chinese feeling--'China for the Chinese and out with the foreigners!' . . . It is, in short, a purely patriotic volunteer movement, and its object is to strengthen China--and for a Chinese program. Its first experience has not been altogether a success as regards the attainment through strength of proposed ends, the rooting up of foreign cults, and the rejection of foreigners; but it is not a failure in respect of the feeler it put out (will volunteering work?) or as an experiment that would test ways and means and guide future choice."[27] These words are worth pondering.

III

The backward nature of the religious superstition of the Boxers inevitably led to their movement's failure, but it does not affect the movement's positive historical character. The Boxers came from many different strata of society and lacked a unified organization. Throughout the movement, no true leadership took shape, but the peasantry is adverse to discipline by its very nature. Peasants have intrinsic weaknesses that cannot be overcome by their own efforts. We can see that the Boxers ' basic organization with the altar as the unit, their leadership by Elder, and their ignorance of the character of modern warfare all meant that they could not succeed against their enemies.

The Boxer movement began through religious superstition, which has no clear class basis. At a time of national calamity, their patriotic calls drew a response from people of many classes. But such a temporary coalition of different social strata and classes could never be more than a loose alliance. As the movement developed and the ranks grew, many bad elements were recruited. Things became even more complex for the Boxers when the Qing dynasty shifted its policy. On June 13, 1900,

> a hooligan named Wang in Tianjin hoisted the Boxers' banner and set up an altar unit in the city's Sanyi Temple to recruit people. Vagrants flocked to join. Upon learning of this, the officials of the neighboring areas called upon Governor Yulu for troops to suppress the Boxers. Their requests were turned down. Moreover, Yulu ordered that the Mutual Protection Bureau (Baojiaju) should send soldiers to protect the altars. After that the influence of the Boxers grew rapidly.
>
> As the news had spread over the city and beyond, one altar after another was established. Within two days, there were more than a dozen in the city and twenty to thirty outside. Each had a membership of from several dozen to several hundred. In the neighboring villages, towns, counties, and prefectures, what had been suppressed before now reappeared.[28]

The Sanyi Temple served as the Boxers' headquarters and assembly point in Tianjin. The establishment of an altar there was the harbinger of the open establishment of Boxer units in the Beijing-Tianjin area.

In Shandong and Zhili, from the winter of 1989 up to May 1900, a serious drought occurred, followed inevitably by popular unrest. That alone brought forward masses of peasants into a mighty antiforeign torrent. When Gangyi and Zhao Shuqiao returned to Beijing from their inspection tour of the Zhuozhou area ordered by the Empress Dowager, the Boxer movement came to its climax in the Beijing-Tianjin area. The imperial court changed its policy from suppression to pacification and decreed that the Boxers be recruited into official militia units.

This meant that the Boxers' antiforeignism received official praise and encouragement, and for the first time the movement really began to look like a "rebellion in response to imperial degrees." Governor Yulu even honored the two Elders with yellow sedan-chairs and a band to escort them. "The rumor spread that the two teachers, Zhang and Cao, had each been granted a yellow jacket and an official hat with two peacock feathers. Afterwards, all the Boxer banners were inscribed with the two characters meaning 'imperially sanctioned.' The Boxers then started to conscript people and collect money in the towns and villages. Anyone who resisted risked the lives of their whole families. Observing that the Boxers could burn and kill with the sanction of the officials, and not seeing through the deceptions, the many common people believed them and joined their units."[29] As numerous "false Boxers" emerged, it became difficult to distinguish between the good and the bad. The movement was getting out of control.

The deviation of Boxers from their original targets (the Christians) was directly related to the Empress Dowager's schemes to use them against the foreigners. But these changes also reflected the serious deficiencies of the movement itself. The Boxer slogan "Uphold the Qing; exterminate the foreigners" is not a logical one, for supporting the Qing could never mean getting rid of the foreigners. In addition, as the movement evolved, the religious superstitions associated these slogans became less and less attractive to the ordinary Boxers.

After entering Beijing, the Boxers set up units in the mansions of the Manchu nobility and dropped their guard against the Qing court. The court, in turn, appointed princes and ministers to command the Boxers and began issuing regulations to control the movement. The Boxers, moreover, lacked a strategy suited to an urban setting. Initially the townsmen gave support to the Boxers and burned incense, worshipped the Boxers' spirits, displayed Boxer charms, and supported Boxer groups with donations of money or food. Yet the "false" units harassed the townspeople by "burning and killing at will," and ultimately the urban residents turned against the Boxers.

Superstitious religious practices, such as possession by spirits and invulnerability to swords and guns, had once helped boost Boxer morale but later became mere tricks the leaders used to deceive the rank-and-file. These same leaders resorted to attributing the deaths in their own Boxer ranks to "lack of proficiency in magic arts" or to the enemy's "heretical magic."

During battle, youths would be put into the front ranks, while the Elders remained in the rear to supervise. These leaders would take to their heels when they saw the vanguard fall. As a result, the ranks "often failed to obey [their leaders] and protested that lives were lost because he lacked magical powers."[30] Once the Boxers began to lose their hold on the minds of the people, any faith in them was lost. It became inevitable that their cause, however great it had been, would end lose its substance quickly.

At the turn of the century, the Chinese national bourgeosie had already come into being. That class should have taken upon itself the task of national salvation. However, it was the peasantry that stood in the van. Even though the peasants were devoid of any leadership themselves and lacked the support of a more advanced class, they were aroused by the righteous cause of nationalism and resolutely shouldered a responsibility that proved too great.

The Boxers shook the world by dealing a heavy blow on the foreigners. This was an outstanding achievement worthy of our respect and admiration. Their backward and even ignorant ways should be acknowledged, without condemning the whole Boxer movement. The fact that such "nationwide madness" occurred at a time when "the people were not sufficiently enlightened" indicates that we should repudiate the backward and ignorant Chinese feudal despotism, but not criticize the peasants for their backwardness and ignorance.

Feudalism imposes a great burden on Chinese history. All the evils and misfortunes in modern Chinese society are rooted in it. In the superstitious practices and antiforeignism of the Boxers we can see the importance and urgency of reform for Chinese society. We can see how expanding productive forces and raising the standards of science and the level of culture are important for the whole nation.

Notes

1. Memorial of Shandong Governor Yuxian, dated Guangxu 25/3/21 (April 30, 1899), in Yihetuan dangan shiliao (Archival Materials on the Boxers, hereafter DASL) (Beijing: Zhonghua shuju, 1959), 1:24.
2. Memorial of Governor Yuxian, 25/11/4 (Dec. 6, 1899), DASL 1:38.

3. Shandong shibao (Shandong Times [Dengzhou]), Guangxu 22/8/5 (Sept. 11, 1896), quoted in Shandong Yihetuan diaocha ziliao xuanbian (Selected Survey Materials on the Boxers in Shandong, hereafter DCZLXB) (Jinan: Qilu shushe, 1980), p. 19.

4. Material from investigations in Chiping county, DCZLXB, p. 192.

5. Report of the prefect at Jinan, Guangxu 25/11/1 (Dec. 3, 1899), in Shandong Yihetuan anjuan (Files Concerning the Boxers in Shandong) (Jinan: Qilu Shushe, 1980), p. 19.

6. Chiping xianzhi (Gazetteer of Chiping County), juan 11.

7. Xinyang xianzhi (Gazetteer of Xinyang County), juan 4.

8. "Report of the Boxer Disturbances against Christians," Huibao 146, Guangxu 25/12/10 (Jan. 10, 1900).

9. Memorial of Shandong Governor Yuan Shikai, Guangxu 26/4/21 (May 19, 1900), DASL 1:93.

10. DASL 1:94.

11. Sawara Tokusake, "Quanshi zaji" (Notes on the Boxer Affair), Qian Bocan et al., eds., Yihetuan (The Boxers, hereafter YHT) (Beijing: Shenzhou guoguangshe, 1951), 1:240.

12. Chai E, "Gengxinjishi" (Records of the Year 1900-1901), YHT 1:305.

13. Tang Yan, "Gengzi Xixing jishi" (Record of a Trip into [China's] Western Regions in 1900), YHT 3:486.

14. "Quanfei jishi" (Records of Boxer Outlaws), in Fan Wenlan, Zhongguo jindaishi (Modern Chinese History) (Beijing: Renmin chubanshe, 1956), 1:346.

15. Memorial of Zhu Zumou, expositor of the Hanlin Academy, Guangxu 25/11/24 (Dec. 26, 1899), DASL 1:43.

16. Memorial of Zhao Shuqiao, minister of punishments and metropolitan prefect, Guangxu 26/5/3 (May 30, 1900), DASL 1:110.

17. Imperial Edicts, Guangxu 26/12/26 (Feb. 14, 1901), DASL 2:945.

18. Guan He, "Quanfei wenjian lu" (Reports on the Boxer Outlaws), YHT 1:470.

19. YHT 2:142, 156.

20. Zhong Fangshi, Gengzi jishi (Diaries of the Year 1900) (Beijing: Kexue chubanshe, 1959), pp. 15, 20.

21. Genzi jishi, p. 11.

22. Lao Naixuan, "Quanan zacun" (Miscellaneous Materials on the Boxers), YHT 4:451, 466.

23. Zhongguo jindaishi ziliao (Materials on Modern Chinese History) 1 (1957): 18.

24. Cheng Ying, ed., Zhongguo jindai fandi fanfengjian lishi keyao xian (Selected Ballads Reflecting the History of Anti-Imperialism and Anti-Feudalism in Modern China), p. 479.

25. Sawara Tokusuke, "Quanluan jiwen" (Reports on the Boxer Disturbances), YHT 1:149.

26. Huang Cunxian, "Nie Jiangjun ge" (Song of General Nie Shicheng), in Renjinglu shicao jianzhu (Comments on Worldly Poetic Fragments), p. 371.

27. Robert Hart, These from the Land of Sinim, pp. 51-52.

28. "Tianjin yiyue ji," YHT 2:141.

29. YHT 2:165.

30. YHT 2:153.

QI QIZHANG

Stages in the Development of the Boxer Movement and Their Characteristics

The Boxer movement had distinct stages, but the developmental process has been overlooked in other studies. A tendency exists to equate a particular stage of the movement's origins with its whole history. Consequently, many issues have become difficult to resolve. Only when we grasp the internal relations of these various stages can we derive some general notions about the Boxer movement. This article approaches the development of the Boxers by dividing the whole movement into three stages, defined and illustrated below.

I

The first stage, including both origin and development, lasted for about twenty months from October 24, 1898, to June 16, 1900. The Boxer movement began in Shandong, where it arose as the people's anti-Christian struggle. After the first Sino-Japanese War this struggle spread steadily throughout the whole province. Those interrelated struggles eventually became a general peasant uprising: the Boxer movement.

The Boxer movement started in the insurrection at the Shibacun exclave of Guan county on October 24, 1898, led by Zhao Sanduo and Yan Shuqin and the so-called Eighteen Stalwarts. Among contemporaries it was well-established that the movement's "originators were the 'Eighteen Stalwarts' of Guan county"[1] and that it "began in the twenty-fourth year (of Guangxu [1898]) with the anti-Christian case in Guan county."[2]

---

*From Yihetuan yundong shi taolun wenji (Collected Articles on the History of the Boxer movement) (Jinan: Qilu shushe, 1982), pp. 118-33. Translated by K. C. Chen with David D. Buck.

That incident had two special characteristics. First, it marked the initial use of the name Boxers (Yihequan) in an uprising. Afterwards this name was adopted by various peasant organizations. The Big Sword Society (Dadaohui) in Qingping county first changed its name to "Yihequan" in the spring of 1899. Before long the Big Swords in many locations were calling themselves Boxers. "Most of the Boxers in Shandong came from transformed Big Sword groups."[3] Zhu Hongdeng's's Spirit Boxers (Shenquan) adopted the name "Yihequan" in May 1899, although "at first only in En and Pingyuan counties, but later spread into Chiping and Yucheng counties."[4] Second, the slogan "Support the Qing and Destroy the Foreigners" (*fu Qing mie yang*) appeared openly on a banner, which predated all other Boxer groups' use of this kind of slogan. From these two points we can tell this uprising differed from previous anti-Christian struggles.

In addition, the Boxers had a distinctive character because of their actions, chief among which was "killing foreigners." At that time the Boxer slogans were "Resist the foreigners' foreign religion" (*dizhi yangren yangjiao*), "Expel foreign officials" (*quzhu yangkou*), and "Bring Christianity to justice and revive China" (*zuona yangjiao, zhenxing Zhongguo*). These indicate a difference from other anti-Christian struggles. For example, the Boxers linked Christianity with foreign officials. This was not plain anti-Christianity because "the Chinese converts take advantage of the common people daily, and the missionaries protect their converts. Consequently, the enmity toward them has become quite deep."[5] This led to plainly self-protective acts of resistance such as the burning of churches. Thus, the Boxers saw the connection between the power of Christianity and imperialist aggression.

Second, they made a connection between "expelling foreign officials" and "reviving China." The Boxers believed that China's decline and poverty were produced by foreigners' activities in China and so pointed out that "only in the past forty years has China become a place where foreigners can operate. If we kill them all in three months, there won't be any more foreigners in North China. The remaining foreigners will flee to their homelands and consequently will have the good fortune to be out of harm's way." The Boxers imagined a day when "Foreigners have been burned, everyone is happy, and there is a bountiful harvest." Thus, the Boxers saw the linkage of national wealth and strength with anti-imperialism. It does not matter whether the Boxers' anti-foreignism was still in the emotional stage and displayed the tendency toward "complete xenophobia," for they added fresh character to the anti-imperialist struggle and were more progressive than other anti-Christian movements.

The Boxers' other important activity was "equal division of grain" (*zhunliang*) to help the poor. This was a new slogan not previously used in anti-Christian struggles. Where did this slogan originate? Since 1894 Shandong had suffered from a series of droughts and floods. By 1898 the famine had affected sixty-one districts. The spring of 1899 "was exceptionally dry and the summer harvest was poor. By the summer pests had eaten the grain in the field and the price of grain had risen sky-high."[6] Because "of several years of harsh famine, the poor have broken up their homes and the corpses of starvation victims are common by the roadside."[7] Many missionaries had cruelly exploited the peasants cruelly, and they took advantage of this situation by enticing their converts "to store up grain in hope of selling at a higher price."[8] Consequently, the starving peasants become more angry. This was one of the reasons for the Boxer uprising.

Thus, the Boxer slogan "divide the grain among the poor" appeared. This slogan had the function of mobilizing the masses, for "when famine victims heard of the division of grain they all joined in happily,"[9] and the struggle quickly spread into southern Zhili. There were two main targets of this slogan. First was the Church itself along with wealthy Christians; second, the wealthy households in general. This first target is revealed by the linking of grain division with attacks on the Christians in slogans from this period, such as "attack the Christians and equally divide the grain," or "famine victims use the excuse of attacking Christians and evenly divide the grain."[10] The latter is reflected in reports such as "They commanded the rich households all to all give grain, but when they didn't get what they wanted, they took it by force."[11]

The slogan "equal division of the grain" applied not only to grain, but also to money, livestock, clothes, and other goods. This aspect of the Boxer movement carried with it opposition to feudal oppression similar to that of earlier peasant uprisings, which called for "Jewelry for us! Livestock for us! Grain for us!"[12] or "Harm to the wealthy and gifts for the poor."[13] However, this aspect has been obscured by the greater attention given to the Boxers' anti-imperialism and thus is often ignored in assessing the movement's character.

In terms of scale and areas of activity we can classify the Boxer movement of this first stage into three subperiods.

1. Origin. [This stage encompasses] October 1898 to April 1899, during which the movement was centered in northwestern Shandong and the border region of Zhili. Zhao Sanduo and Yan Shuqin started the rebellion in the "Shibancun exclave of Guan county," which was an old designation for a piece of territory that comprised twenty-four villages and was a part of Shandong's Guan county, even though it was located within Zhili. Such administrative anomalies were call "exclaves" (*feidi*). This land was 140 *li* from the Guan county capital and surrounded by Nangong, Wei, Qihe, Qu, and other Zhili counties. Within that special piece of Guan county were one large church and nine chapels, and "as they grew steadily, the hostility deepened between the Christians and the non-Christians."[14]

Zhao Sanduo was the master of the Plum Flower Boxers (Meihuaquan), and Yan Shuqin led the Red Boxers (Hongquan). Both groups had a popular base of support and practical experience in struggle. All these conditions enabled the "Shibacun exclave" to become the birthplace of the Boxer movement. After the uprising began, "Zhao Sanduo rose to be its leader and assembled several thousands. Soon it spread across several provinces causing great consternation by its rapid growth."[15] Once it began, the Boxer uprising's rapid growth and huge scale exceeded previous anti-Christian struggles, thereby shaking the Qing dynasty's authority. Through suppression by the army, the rebellion was curbed temporarily. Zhao Sanduo shifted his activities to Wuyi, Cangzhou, and Zhengding counties in Zhili, while Yan Shuqin remained active in the Shandong-Zhili border areas.

2. Development. From April 1899 to January 1900, the center of activity shifted into Shandong province. Zhu Hongdeng, the leader of the Spirit Boxers, moved around in Chiping county and started to spread boxing there. "It was a time of famine in inland Shandong, and the poor lacked food." "Masses of them banded together into groups of thousands."[16] Among the 860 villages in Chiping, more than 800 had groups practicing boxing. In May 1900, Zhu Hongdeng changed the Spirit Boxers' name to Boxers (Yihequan) and began operating around En and Pingyuan counties. Before long they had spread "the practice of their groups everywhere between Jinan and Dezhou."[17] Meanwhile, many Big Sword groups in other areas also adopted the name "Yihequan." In less than half a year, the Boxers had spread to most parts of the province.

3. Advance. From January to June 1900, the center of activity shifted to Zhili. The Boxers had spread generally through Zhili during the winter, and by the spring of 1900 they had reached the environs of Beijing. On May 22 the Boxers (Yihetuan) ambushed Qing units at Shitingcun in Dingxing county, killing General Yang Futong. On May 27 they damaged the railway, dismantling a section between the Liuli River and Zhuozhou. On the 29th, the Boxers "occupied Zhuozhou city, hung banners from the gates, and established strict security, even searching the residents as they passed the gates."[18] On June 5 the Boxers carried out an assault on Qing units at Gaobeidian and defeated two companies, which had been sent to reinforce that position. At that time, in Zhili, "in the northern section Zhuozhou, Xindian, Dingxing, and Ansu and the southern counties of Zhangdeng . . . were all controlled by the Boxers, who numbered some twenty thousand, or at least several thousands."[19] Around Beijing, the Boxers already had acquired explosive force.

What was the relationship between the Boxers and the Qing government at this juncture? The dominant view at present holds that the slogan "support the Qing" was the Boxers' guiding principle. However, there is no evidence to support this view. First, the Boxers gave priority at this time to "destroying the foreigners." Their chief target had become the power of Christianity, which indicated that nationalist contradictions had become the chief contradiction within Chinese society. When the Boxers fought with government troops, it was a result of Qing protection for Christian churches. Consequently, the Boxers' "support for the Qing" was subordinate to "destroying the foreigners." If the Qing government opposed "destroying the foreigners" it also became an object of attack, and there was no talk of "supporting the Qing." Second, although the Boxers did not often attack the Qing dynasty at this time, this cannot be taken as proof that they supported the Qing. The armed forces of the Boxers and the Qing opposed one another, and both followed primarily defensive approaches, which only indicates that the two sides' strength was roughly equal. From this comparison it can be seen that the Boxers did not have overwhelming strength behind their attacks. Still, this does not prove that they were unable to attack, for they did so in such places as Gaobeidian. Besides, there are many later examples of the Boxers attacking. Finally there was a wide variety of slogans with different meanings, such as "bring the Christians to justice and revive China," "protect China and expel the foreigners," "protect the homeland and expel the foreign officials," "destroy the Manchus and restore the Chinese," "oppose the Qing and support the Ming; drive out the foreigners," "The Qing declines and China is strengthened," and "kill the foreign devils and create trouble for the Qing." All of these make it hard to accept that the Boxers' guiding principle was support of the Qing.

At that time there were people who said that the Boxers were using the slogan "support the Qing and destroy the foreigners" in order to "flatter the Court."[20] Others have pointed out that when the Boxer uprisings began, "The Boxers worried about official intervention against them on grounds that they were an antidynastic conspiracy, so they created the slogan 'support the Qing and destroy the foreigners' to draw the people's support while hoping to placate the officials. Thus they distinguished themselves from the religious sects such as the White Lotus and Heavenly Gate (Tianmen)."[21] This analysis is convincing for it explains the Boxers' support of the Qing as a means of protecting themselves while carrying out the policy of attacking the foreigners.

The support for the Qing influenced the government's policy toward the Boxers to a certain extent. Some officials suggested, "Either extermination (*jiao*) or pacification (*fu*), and sometimes a policy combining both techniques, have been the traditional means of quelling uprisings."[22] Extermination and pacification have been the two antirevolutionary means of feudal rulers for keeping the people down. Sometimes both means were employed; at other times the two policies were used in alternation. Prior to the outbreak of the Boxer movement in 1896, Shandong Governor Li Bingheng adopted a combination of extermination and pacification toward the Big Sword uprising led by Liu Shiduan. In 1898 Governor Zhang Rumei carried out a policy of strict suppression against the uprising of Tong Zhenqing in the Shandong-Zhili border area.

Why were these two cases treated differently? In case of the Liu Shiduan's Big Swords, "They did not intend to create disorder," but the uprising occurred "Because of the hatred between the people and the Christians."[23] The Tong Zhenqing uprising, however, was a "planned antidynastic conspiracy and a crime of great proportions."[24] The Boxers' "support the Qing and destroy the foreigners" actually was a slogan intended to clarify that they were not planning a conspiracy against the dynasty.

Many officials adopted a somewhat sympathetic attitude toward the Boxers. Some reported, "In Shandong the Boxers did not want to create disorder, but they are bullied daily by the Christians, so that hatred penetrated their bones . . . and with no way to obtain justice and no means of protecting themselves, they had to organize their own militia in order to protect themselves and their families."[25] At the same time, other officials worried that too harsh suppression of the Boxers might make the people "turn to extremes," "explode," or "provide an excuse for intervention by foreign troops." Thus extermination "among the available alternatives may be even more disastrous."[26] Therefore, from the beginning, the Qing government adopted policies combining extermination and pacification. Implementation of this approach had two important aspects. First, the Boxers "were to be dispersed, but if they attack our officials or troops, we order stubborn resistance and a firm display of our martial power."[27] Second, the troops were used only to back up the pacification policy, for the government "arrested the leaders and dispersed the followers"[28] and thus achieved the goal of ending the disturbances. For example, leaders such as Yao Wenqi, Zhu Hongdeng, and Monk Benming all were executed under this policy. Thus the policy of combining extermination and pacification may have appeared different from that of strict suppression but had the same goal.

Zhang Rumei and Yuxian both had suggested policies combining extermination and pacification, but with Yuan Shikai the situation was different. His intent was to suppress the Boxers with force; however, he was under instructions from the Court "not to suppress [the Boxers] indiscriminately, thereby forcing them to extremes and producing an even greater disaster."[29] He was instructed to restrain himself and not alter the policy of combining extermination with pacification.

After November 1899 a change in policy occurred. The Empress Dowager commented in her edict of November 21: "Now the situation is becoming worse daily. The foreign Powers eye us covetously and jostle to be the first to attack, but considering that our financial and military strength is deficient, we cannot initiate the quarrel. Once the situation changes and we are pressed, we must resist forcefully. I will not consent to defeat. The only proper thing is to show our martial spirit and show we share a hatred toward the enemy. Whether we will be victorious or defeated cannot be predicted."[30] How could the Empress Dowager, who feared the foreigners as if they were tigers, come to show such bravery? Some observers recorded, "After the Empress Dowager foiled the [Restoration] plot, she refused to see the foreign representatives, and her hatred of the foreigners became intense. The Boxers, using the actions of the Christian converts, began proclaiming 'destroy the foreigners.' With the support of officials who approved of this tactic, the Empress Dowager followed their wish to use the Boxers to destroy the foreigners."[31]

In other words, the Empress Dowager found that the Boxers could be useful. Thus on January 11, 1900, an edict further confirmed that the Boxers "practiced martial arts to protect themselves and their families with the purpose of mutual protection and support" and directed, "only determine if they are outlaws or not. Pay no attention if they are secret societies (*hui*) or religious sects (*jiao*)."³² Actually, this was an attempt to recognize the Boxers as a legal group and preparation for a policy that would protect them. Before long the Imperial Censor Zheng Binglin suggested that the Boxers be protected "by having officers train their privately organized groups." However, due to the objections of officials such as Prince Yikuang, Governor General Yulu of Zhili, Shandong Governor Yuan Shikai, and Censor Li Zheying, the change was shelved temporarily. Consequently, the Qing government's policy toward the Boxers in this period did not change from the previous strategy of combining suppression with pacification.

II

The second stage, which lasted only for two months, from June 16 to August 14, 1900, was the movement's high tide. June 1900 was a great turning point in the Boxer movement. Prior to this time the Qing government had followed the policy that combined suppression and pacification. Because the foreign Powers insisted on sending troops to Beijing to protect their embassies, the situation there and in Tianjin became tense and fighting was on the verge of breaking out. The gang around the Empress Dowager, feeling their own control to be in great danger, immediately changed Qing policy to simple protection for the Boxers, intending to use them to provide security for themselves. Zhao Shuqiao, minister of the Board of Punishments and prefect of the Metropolitan Prefecture of Beijing, wrote to the Empress Dowager advocating once again the policy of protection for the Boxers: "The Boxers have steadily grown and we cannot kill them all. A better strategy is to protect them and use them by controlling them with regular military officers and enrolling them into the ranks. Because of their hatred for Christianity, they will fight courageously."³³ The Empress Dowager had already decided to protect the Boxers, but because of the official opposition she had yet to carry through her decision. In early June, she sent Zhao Shuqiao and Gangyi, a grand councillor and assistant grand secretary, to Liangxiang,

Zhuozhou, and other places in Zhili to placate the Boxers. Thereupon, most officials accepted the Court's policy. The librarian of the Supervisorate of Imperial Instruction, Tan Ji, proposed to gather the Boxers "into a special army, provide them food, and discipline them according to military rules."[34] Gangyi stated "there is no reason to suppress" the Boxers. On June 16 the Empress Dowager summoned an imperial audience for the high officials and issued an edict stating that from among the militia (*tuanmin*), "the young and strong should be selected and enrolled into an army."[35] In this fashion the policy of protection came into existence and the Boxer movement entered a new stage.

In this new stage the Boxers came to Beijing and Tianjin in great numbers, and that region became the center of Boxer activities. With the approval of the Empress Dowager the Boxers had obtained legitimacy, and this had both positive and negative effects on their development. On the positive side this promoted the growth of the Boxers and quickly took the movement to its high tide. In the spring of 1900 the Boxers appeared in Beijing, but their numbers were few. Once the Qing announced their policy of protection, "Boxers from the surrounding areas streamed into the city day and night." According to figures in this period of greatest activity, "there were more than 800 altars (*tan*), and if each had 100 members, there were 80,000 Boxers."[36] The official estimates were "not less than 100,000."[37]

It can be seen at a glance that the Boxers developed quickly. At this time, not only in Shandong and Zhili, but north to Mongolia and the Northeast, and in Sichuan in the southwest, the Boxer organization underwent a great expansion. On the negative side, however, the Boxers who entered the Beijing-Tianjin region became subordinated to the Empress Dowager and her gang.

The Qing government adopted a whole set of procedures to turn the Boxers into a tool for their own ends. First, members of the imperial family and high officials were given command over the Boxers. On June 23 Imperial Censor Liu Jiamo noted that "the Boxer units are disorganized and people are not under any general command," and he proposed that the Court "specially depute a Prince and a high official to become the officers reviewing these militia units on the principle of 'official oversight of the people's management' (*guandu minban*)."[38] This advice was adopted quickly, for the Empress Dowager promptly appointed Prince Zhuang [Zaixun] and Assistant Grand Secretary Gangyi to command the Boxers, and Yingnian and Zailan from the Imperial Guard units to be their assistants. Second, to control the Boxers more strictly, they established a registration system for Boxer groups at the Princes' Household Office in which "the registered groups are official militia and carry the designation 'approved Boxer Unit,' while the unregistered are private militia and do not carry the 'approved' designation."[39] Third, they published "Regulations for Boxer Militia," requiring "the Boxers and the army to cooperate as if they were one family," "to respect orders," and to follow military regulations." Fourth, those Boxers who failed "to obey orders were labeled "false militia" and were designated as "wrongly claiming the approved Boxers rights" and would be dealt with "by applying the laws on banditry" to them. In this manner most of the Boxer groups entering Beijing were turned into pawns of the Qing government.

However, outside the capital region there were many areas where this policy of protection was not followed. Many local Qing officials saw clearly that protection of the Boxers was merely an expedient policy for the Court. Even the Empress Dowager herself had to recognize, "Suppression would only increase the disaster. The only way is to try to subordinate them and thus solve this situation gradually." She also stated, "This is an inescapable and embarrassing difficulty."[40] This phase of protection could not be sustained for long.

Because of this some local officials, although they did not dare openly to reject imperial direction, in practice did not carry out the policy of protection. Shandong Governor Yuan Shikai is an example. On June 16 when the Qing Court issued its command for protection, he used this as an excuse to countermand his order to suppress the Boxers. Yuan Shikai published an announcement that all the Boxers within Shandong should "go north to help in the battle." In this document he also stated, "At present, foreign troops are assembling around

Tianjin, and Boxers should all join the battle against the enemy; those who remain in the interior are bandits pretending to be doing good . . . but we will deal with them severely as if they were bandits. If they dare to resist, we will kill them without hesitation."[41] In reality he switched to a policy of complete suppression of the Boxers. Yuan Shikai thus carried out the large-scale slaughter of the Boxers in the name of pacifying "bandits pretending to be Boxers." For example, on July 4 at Haifeng, more than 100 Boxers were killed; on August 5 at Yangxin, more than 500 were slaughtered; and the following day, in Ling County, more than 140 Boxers died. Yuan's hands were soaked in Boxer blood, and he carried out a serious crime against the people.

At this time there were many local Boxer groups that refused to accept the Qing amnesty and enrollment. Especially some of the older Boxers from the Shandong-Zhili border region continued the struggle uninterrupted. Zhao Sendou and Yan Shuqin, having stirred up southern Zhili, returned to action in Shandong. In the two districts of Yanshan and Qingyun in southern Zhili more than 20,000 Boxers "assembled when needed but otherwise were dispersed in their villages." "Gathering the scattering . . . they are everywhere and they expand so rapidly it is hard to estimate."[42] The situation was becoming more menacing to the Qing dynasty. Various county seats such as Qingyun in Zhili and Yangxin in Shandong were occupied by Boxers. They had coordinated groups involving several counties to oppose Qing military suppression. For example, Liuzhongkou village in Zhanhua county was "the general [Boxer] headquarters for five counties"--Binzhou, Lijin, Yangxin, Putai, and Zhanhua. Boxers for the six counties of Qihe, Huimin, Shanghe, Zouping, and Zhangqui made the Boxer Sun Yulong (a.k.a. Sun Yutong) their grand elder and master (*zongguan dashixiong*).

From this it can be seen how the Qing dynasty was thrown off balance and its authority was failing. Although there was a phase of enrollment and protection, this policy was limited to certain areas, while in many locations a policy of extermination was carried out in the name of protection. Some Boxers were enrolled as official troops, but the great majority never were.

Those who believe that the Qing policy of protecting the Boxers was completely implemented or that the Boxers readily accepted enrollment are incorrect.

At this stage the Boxers had two obvious characteristics. First, they had split over the issue of "supporting the Qing," and two tendencies can be discerned from the patterns of Boxer actions. In one category were those who supported the Qing; in the other were those who never emphasized support for the Qing. The first element was composed of most of the Boxers in the capital. The character of the Boxers underwent a great change which was especially obvious among the Boxers in the Beijing-Tianjin region. "Vagabonds and bandits swaggering about and causing trouble all set up groups. The momentum behind their sect was tremendous. Even the honest and wealthy had to join for self-preservation. Their elders claimed to be divine and and spoke of 'Heavenly Rights.' "[43] Because of their popularity they could be insufferably arrogant, "and they asserted their authority more than did local officials." One observer pointed out, "There are some who joined out of hope of booty, those who joined hoping for revenge, those on the edge of destitution who joined to avoid starvation, and rich people who established their own Boxer units for self-protection and fear of the Boxers' depredations."[44]

Especially important were the nobility and officials who set up Boxer units such as the son of the Manchu Board of Presidents who made himself chief of some Boxers; the son of Jingshan, a retired high official, who became a leader; the Manchu Prince Zailian, who permitted Captain Wenshun and a hundred Boxers to live in his outer garden; and the eldest Princess Yizhun, who allowed two hundred Boxers to camp at her palace. Even units of the Imperial Guards (*hushenying*) commanded by the Manchu Prince Zaiyi established a Boxer altar in the vacant land of his palace grounds, making the connection between the Boxers and the Imperial Army complete.

This was the phase in which "everyone from princes and ministers down to common people joined the Boxers." Without question these Boxers were duped by the Empress Dowager's gang. They had become part of the Court's internal power struggles under the direct command of Prince Zaiyi. These

Boxers certainly did support the Qing, but this really meant the Empress Dowager's interests.

Second, there were those Boxers who supported the Guangxu emperor's interests, but they did not occupy controlling positions. It was they who proclaimed, "Our emperor will resume power soon and the Boxers will be his loyal ministers" and "Protect the Qing from misfortune; the emperor must come to no harm." This second variety of Boxers supported the legitimate dynastic authority, meaning the Guangxu emperor and the Empress Dowager.

Third, some Boxers claimed to support the Qing, but actually did not. Outside of the Beijing-Tianjin region this last variety was widespread during this stage. As stated above, they were never enrolled under Qing colors and continued to battle the foreigners as well as Qing troops. Their political program did not support the Qing.

When we talk of "Support for the Qing" in this stage of the movement, concrete distinctions are required. In regard to the first two varieties in the Beijing-Tianjin region described above, they supported the Qing in word and deed. Their position was "kill all the foreign devils," by which the Qing would achieve unity and peace. They put support for the Qing ahead of killing foreigners, but that fell short of being a viable program. As for the third variety of Boxers still in the rural areas, their support for the Qing was only verbal, and their character remained essentially the same as in the first stage: an organization intended to accomplish the goal of "Destroying the foreigners."

A second special characteristic of this stage was the change in Qing policy to one of protection for the Boxers. Some Boxers never enrolled under that protection but took advantage of it to expand their strength and attack the Qing army. While such attacks were not common in the first stage, during this high tide they became common. For example, the Boxers in the southern part of Beijing's environs frequently attacked Qing units in the period. On June 18, 1900, at the battle of Bazhou, the commander of the Left Division of the Guards' Army, Yang Mushi, was leading three battalions northward under orders. When they stopped to rest outside Bazhou city, the Boxers burst forth from the city gates and "the suburban villages sent groups who coalesced into a mob which descended waving swords and guns, killing many" Qing troops. Even Yang Mushi was almost captured, but he managed to escape.[45] On June 19, at the battle of Laishui, while a Qing army unit was fording the Lai river, their supply detachment was billeted in the village of Zhangzhuang. Boxers took advantage of darkness and attacked at night, "scattering the soldiers and capturing the ammunition." The main unit responded quickly, "but the Boxers' strength was too great and they killed about eighty soldiers while suffering only a few wounded themselves."[46] On July 13, at the battle of Leling, the Boxers attacked the church and foreign compound at Zhujiazhai with the knowledge that Qing troops were protecting it. The Boxers split their forces into three. One unit appeared in front of the village and feigned an attack, drawing out the Qing cavalry, while the second broke into the back of the walled compound and set fire to three foreign buildings. As the Qing cavalry emerged to take on the first Boxer unit, the third was lying in ambush. When the Boxers broke off their successful engagement, the Qing troops "returned to the compound, but could not extinguish the fires."[47] At Dezhou, on July 26, Boxers swooped down on the city. "Disregarding danger, they attacked from all directions," surrounding two Qing battalions and killing five officers and more than seventy soldiers, "thus winning a major victory."[48] In these battles, the Boxers attacked even when not assured of victory and used a variety of battle tatics, showing that those who claim they were incapable of conducting real military attacks are wrong.

As mentioned, we cannot be absolutely certain whether the Boxers supported or opposed the Qing. Overall it is obvious some Boxers accepted Qing enrollment and protection, but we cannot escape the fact that most did not. Those who insist that either of these two positions is completely correct simply do not understand the historical situation.

If we compare the first two stages, we can see that the content of the Boxer movement had changed. The major proofs of this change are threefold. (1) In the first stage, the Boxers' chief actions were attacks against Christians. While the Qing made attempts to suppress them, the Boxers themselves seldom initiated attacks on the government. In the second stage, the Boxers themselves undertook frequent attacks on the Qing. Thus, the anti-Qing tendencies of this second stage became clear. (2) In the first stage, a secondary Boxer activity was the division of grain; this was expanded even further in the second stage. Thus in the prefectures of Kaizhou, Nanle, and Qing-feng, and in the adjoining counties of Daming and Yuancheng in Zhili, there were instances of "masses dividing the grain." In Shandong, in addition to carrying on the struggle to distribute food to the poor, the Boxers also broadened the scope of their struggle by "requiring each village to provide provisions," and seeking "food loans" from the officials.[49] Thus the Boxers' antifeudal character increased at this stage. (3) In the first stage, the Boxers used the slogan "support the Qing," but they did not carry it out, while in the second stage, a three-way split developed within the movement over this matter. One group continued not to implement the slogan, while another made "destroying the foreigner" into the means by which it protected the Qing. In general, the Boxers split into two elements. One tended toward opposition to the Qing and the other moved toward full support. Those comprise the basic dis-similiarities between the two stages.

III

The third stage of decline lasted from August 14, 1900, through 1902, or about two years. On August 14, when the Allied armies representing eight nations occupied Beijing and the Empress Dowager fled to Xian with the emperor, the Boxer movement entered the stage of decline. On September 7, the Qing government proclaimed, "This affair arose because the Boxers created trouble. Now we intend to root out the difficulties, but this cannot be accomplished without pain."[50] Not long before, the Empress Dowager had offered high praise for the Boxers, saying, "They have not cost the state a single soldier, nor a single peck of grain. Even teenagers have taken up arms to protect our society."[51] In this third stage, she changed her tune and openly brandished a sword to murder the Boxers. Thus the Qing policy underwent a complete about-face from protection to extermination. Yet even under the combined pressure from the foreigners and the Qing, the Boxers did not surrender or cower, but continued their struggle.

The third stage of the Boxer movement can be divided into two subperiods.

1. From August 1900 to September 1901 was the period of Boxer resistance to Qing and increased foreign pressure. After the Allied armies had occupied Beijing, they expanded the area of their aggression and brutally suppressed the Boxers. At many places the Boxers conducted raids on the invading troops. A unit of Americans stationed in Beijing was "suddenly attacked" by Boxers and had to be rescued by "Indian troops." A German unit in Beijing also was raided by Boxers, and another detachment of German troops protecting the Luanzhou railway bridge as well as the Russian and German troops at Shanhaiguan were attacked by the Boxers.[52] In southwestern Laishui county, one group of Boxers, numbering 3,000 to 4,000, "divided into three groups and looted all the rich households." When French troops were sent out from Baoding to suppress them, they adopted tactics of wearing down the enemy. They waited until the French troops were nearby, then fled into the hills without a trace. Once the French returned to Baoding, the Boxers came back from the mountains and raided as before.[53]

The Boxers blocked the Allied armies' deepening aggression at many locations. When German marines and Indian troops attacked Liangxiang, the Qing troops fled, but the local Boxers bravely counterattacked, causing many enemy casualties. As the invading armies were attacking Baoding, the Qing units, as usual, scattered, while a group of Boxers "wearing black headwraps bravely went forth to meet the enemy" and inflicted heavy casualties on the foreigners.[54]

Concurrently, the Boxers were counterattacking against the Qing policy of extermination. In the area around Baoding, and including Maozhou, "The Boxers are extremely fierce. The official troops have carried out suppression on several occasions, but they will not remain pacified" and "reappear to resist the soldiers." A unit of Tianjin gunboats arrived in Bazhou to suppress the Boxers, but they resisted so fiercely that "all the gunboats were dispersed and many sailors were wounded."[55]

2. From September 1901 through 1902 was a period of revived Boxer uprisings. On September 7, 1901, the Qing government signed a group of unequal treaties, the Protocol of 1901,
which contained unprecedented national humiliation and betrayal. This protocol provided that China would pay to the
eleven invading foreign nations a total of 450 million silver
taels over the next thirty-nine years. The provinces all shared
in the indemnities and had to establish new bloodsucking levies
and taxes. These led to a rebirth of the Boxer struggle. In this
period of revived Boxer activity, two uprisings had the greatest
influence. One occurred at Julu on April 23, 1902, and involved Boxers led by Zhao Sanduo and a united village association (*lianzhuanghui*) from Guangzong county led by Jing
Tingbin. The other was the Ziyang uprising in Sichuan, also in
April, led by Li Gangzhong. Both of these uprisings had a
profound influence in promoting the revolutionary situation
during this third stage.

The Julu uprising sometimes is referred to as if it were led
by Jing Tingbin, but actually it was the Boxer remnants led by
Zhao Sanduo who were most important. Records indicate that
among the Boxers was a unit from Wei county displaying "a
black flag with a white border, led by one 'Old Red Zhao'
(Zhao Lao zhu), who previously had served as a Boxer leader"
and "now had rejoined the Boxers seeking revenge."[56] Behind
this nickname "Old Red Zhao" was the same man, Zhao Sanduo, who had led the Boxers in this region earlier. He was not
willing to participate in just any uprising, but "took the lead in
organizing this one."[57] Thus this instance can be directly linked
with Zhao Sanduo. After this outbreak at Julu, the banner
"abolish the Qing and destroy the foreigners" (*sao Qing mie
yang*) appeared, and officials said, "the traitors of the region
all support them." "Gathering like a swarm of insects," they
had "within a few days fanned the flames of disorder."[58] Thus
its great importance is obvious. "The Ziyang uprising used the
slogan "destroy the Qing and exterminate the foreigners" (*mie
Qing jiao yang*). They "began by attacking the churches and
killing the Christians. Then in the name of robbing the rich to
help the poor, they grabbed everything and wantonly killed
without regard to whether people were Christians or not."[59]

The slogans "destroy the Qing and exterminate the foreigners" and "attack the rich and give to the poor" made a deep impression. Those who practiced boxing in various counties responded, and their bands gathered near Chengdu. There was a popular song that went, "Boxers approach the outskirts, clamoring to attack the churches. The people are abuzz, while the officials are alarmed; the people watch and talk with drawn faces. No matter what strategy is proposed, they are frightened."[60] Thus we can see that the Qing dynasty's control in Sichuan already was crumbling and tottering.

In the last half of 1902, the Boxer uprisings in various locations were suppressed, thus marking the conclusion of China's last major old-style peasant war. Still, the Boxers did not completely cease their struggle. Remnants remained active in many localities. The Loyal Army (Zhongyijun) struggled both to resist Russia and to oppose Qing surrender, and thus continued the Boxer movement in the Northeast (Manchuria). By the spring of 1903, the resistance had been suppressed by Qing and foreign forces. The Boxers again rebelled that same summer at Tuoer mountain in Shandong's Rizhao county under the leadership of Li Yongjiu. After that, the activities of the Boxers are not fully recorded.

In 1905, on the Sichuan-Hunan border, there was a Boxer uprising led by "Ropemaker" Ding which proclaimed "destroy the Christians; restore the Boxers" (*damie yangjiao chongxing quanhui*). At the same time throughout Sichuan, "People were unsettled and anxious everywhere because placards opposing foreigners and Christians were widely displayed."[61] In 1906, at Zuoyun county in Shanxi, Boxers assembled to drill in remote sites and then marched en masse on the cities, demanding that officials distribute grain to them. In 1907, at Qu county in Sichuan, Boxers recruited teenagers and drilled day and night; in Kai county, the Boxers carried out a razzia, "bringing harm to wealthy households of the area."[62] In 1909, Su Zilin and Sun Gaoru planned to set fire to official buildings and churches. In 1911, when Sichuan established the Association to Protect Railways, various armed units arose and attacked Chengdu. A great many Boxers participated in this, but only in a sporadic or individual manner, so they had little effect. The Boxer movement as a large-scale, traditional-style peasant war basically was over by 1902.

What was the special characteristic of the Boxers at this stage? First, in the face of the joint Qing and foreign suppression, as the reactionary nature of the Qing was exposed, the Boxers directed their main attacks at the Qing dynasty itself. After the Allied armies occupied Beijing, the Boxers of the Shandong-Zhili border region fought even more fiercely in a struggle to the death against the Qing. On September 24, 1900, at the battle of the Jade Emperor Temple in Jiyang, the Boxers, "encircling from three sides, charged forward against the Qing troops," who "attacked from all sides but could not repulse the enemy" and so fled in defeat. In pursuit, Expectant Magistrate Cha Rongsui was cut down from his horse.[63] In mid-October, at the battle of Ansu in Zhili, a Qing unit "came upon a Shaanxi Boxer unit on the road," and the Boxers "without any premeditation attacked, leaving both sides in disarray," but inflicting heavy casualties on the Qing troops, with 13 officers and 763 men killed or wounded. Also in that month, Boxers at Miyun broke into the walled county seat. "They killed the magistrate and released all the prisoners in the jail."[64] This kind of killing of officials and military officers was uncommon in the first two stages. By the time of the uprisings in Julu and Ziyang, everyone knew that the Boxer struggle was aimed primarily at the Qing dynasty.

Second, in connection with the change of the primary target in their struggle, the Boxers issued new slogans replacing their original ones. In the first stage, the Boxers' public slogan was "support the Qing, destroy the foreigners," but in private it was "oppose the Qing and support the Ming." In the second stage, although the slogan "oppose the Qing, restore the Ming" continued to circulate secretly, the public slogan of "support the Qing and destroy the Foreigners" indicated the dominant trend. In both the first two stages "destroying the foreigners" was the main Boxer goal. Yet, in the third stage it was different. Not long after the Allied armies occupied Beijing, the Boxers around Xi, Xaoyi, and Shilou counties were the first to change their slogan to "restore China and destroy the foreigners" (*xing Zhong mie yang*), discarding the element of support for the Qing. In June 1901, the Sichuan Boxers used the slogan "destroy the Qing, exterminate the foreigners." That same year in Shenzhou, Anping, and other parts of Zhili, Boxers used the slogan "eliminate the Qing and destroy the foreigners," while in August at Xiong and Ba counties in Zhili the slogan "oppose the Qing and destroy the foreigners" appeared. The change in these slogans reveals that the Boxers had begun to realize the connections between eliminating the Qing and destroying the foreigners. This can be taken as an historical turning point in the Boxer movement.

Third, in comparison with the earlier two stages, this third stage had clear tendencies toward democratic revolution. At this stage, the switch from support of to opposition to the Qing constitutes a qualitive change in consciousness. Futhermore, the Boxers linked up opposition to the Qing with antiforeignism and shouldered the dual burden of opposing imperialism and feudalism. It is not simply chance that the Boxer movement falls between the 1898 Reform Movement and the bourgeois democratic revolution of 1911. Sun Yat-sen has pointed out that prior to the Boxer movement, Chinese public opinion viewed his revolutionaries as "thieves and scoundrels, traitors and heretics, and we were reviled in an endless stream of invective" or "seen as poisonous snakes and wild animals." After the Boxers that opinion changed, "and among men I know, more and more came to me, took my arm, and said sympatheticly, 'it was a shame they didn't succeed.' The change was like the leap from one side of a chasm to the other."[65] Lu Xun said, "The Reform Movement of 1898 failed, and then following the Boxer Upheaval in 1900, everyone knew the government was not capable of ruling and wished to overturn it."[66]

Why did such a change occur? It was because the Boxers attacked and weakened the feudal rule of the Qing dynasty, revealing the Qing ruling circle's rottenness and incapacities. They destroyed the facade of Qing authority, thereby advancing the awakening of the people and the development of the national bourgeois democratic revolution. Thus, this third stage of the Boxer movement was not led by the bourgeoisie, but rather shows the peasantry's major role in China's democratic revolution. It revealed the trend of the Boxer struggle to join in the bourgeois democratic revolution. For example, in 1906, the Boxer leader He Rudao in southern Sichuan linked up with Revolutionary Alliance (Tongmenghui) member Li Shi to start the Jiangyou uprising. In 1911, the Boxers in Sichuan joined in the revolution led by the bourgeoisie.

In 1908 Lenin had written, "The revolutionary movement in China opposing the medieval system has, in recent months, displayed unusual strength. Indeed, we cannot make a clear evaluation of this movement because of limited information, yet we do know a great deal about rebellions occurring all over China. So there is no doubt that 'the new spirit' and 'European ideas' have grown there, especially following the Russo-Japanese War. Therefore, the old-style riots of China can turn into conscious democratic revolution."[67] What he meant by "rebellions occurring all over China" and "old-style riots" included the Boxer movement. Examination of historical facts shows that his comments were correct.

In conclusion, the problems of research on the Boxer movement are exceptionally complex. To oversimplify these is incorrect. After examining the whole process of its history, I have tried to divide the Boxer movement into three stages. My attempt was directed at understanding its nature, its characteristics, and its transformations from a general perspective. Thus, we can come to know the real issues and, through scholarly discussion, reach an appropriate evaluation.

Notes

1. Huibao, no. 153.
1. Lao Naixuan, "Gengzi fengjin Yihequan huilu" (Collected Records on the Imperial Prohibitions of the Harmony Boxers in 1900), in Qian Bocan et al., eds. Yihetuan (The Boxers, hereafter YHT) (Beijing: Shenzhou guoguangshe, 1951), 4:459.
3. Huibao, no. 187.
4. Huibao, no. 146.
5. "Shandong xunfu Li Bingheng zhe" (Memorial of Shandong Governor Li Bingheng), Guangxu 22/6/24 (Aug. 3, 1896); Yihequan dangan shiliao (Archival Materials on the Boxers; hereafter DASL) (Beijing: Zhonghua shuju, 1959), 1:6.
6. Qing Dezong shilu (Veritable Records of Emperor Guangxu's Reign), juan 450.
7. "Yancheng xian ju Yizhoufu bing" (Report from Yangcheng county to Yizhou prefecture), ms.

8. "Yushi Gao Xizhe zhe" (Memorial of Censor Gao Xizhe), Guangxu 25/12/5 (Jan. 5, 1900), DASL 1:48.
9. "Yangcheng xian ju Yizhoufu bing."
10. Ibid.
11. Shandong Yihetuan anjuan (Documents on the Boxers in Shandong) (Jinan: Qilu shushe, 1980), 2:926.
12. Tang Shisan xiansheng ji (Collected Works of Tang Shisan), jian 20.
13. Ding Yaokang, "Baocun canye shi houren cunji" (Remnants from Scattered Records for Our Descendants), in Chujie jilue (Records of Surviving the Catastrophe).
14. SDAJ 1:447.
15. Guanxian zhi (Gazetteer of Guan County), juan 10.
16. Li Di, "Quanfei huoguo ji" (Accounts of the Disasters Committed by the Boxer Bandits), p. 6.
17. Huibao, no. 136.
18. Lin Xuecheng, "Zhidong jiaofei diancun" (Collected Telegrams Concerning the Suppression of the Bandits in Zhili and Shandong), in Yihetuan yundong shiliao congbian (Collected Materials on the Boxer Movement; hereafter SLCB) (Beijing: Zhonghua shuju, 1964), part 2, p. 137.
19. Ibid.
20. Hu Sijing, "Lubeiji" (Collected Writings of an Itinerant Secretary), in YHT 2:483.
21. Wu Yong, "Gengzi xishou congtan" (Collected Essays Concerning the Westward Flight of the Emperor in 1900), in YHT 3:373.
22. "Yushi Zheng Binglin zhe" (Memorial of Censor Zheng Binglin), Guangxu 26/5/23 (June 19, 1900), DASL 1:156.
23. "Memorial of Shandong Governor Li Bingheng," Guangxu 22/6/24 (Aug. 3, 1896, DASL 1:5.
24. "Shandong xunfu Zhang Rumei pian" (Memorial of Shandong Governor Zhang Rumei), Guangxu 24/7/29 (Dec. 5, 1898), DASL 1:18.
25. "Yushi Huang Guijun zhe" (Memorial of Censor Huang Guijun), Guangxu 25/22/25 (Dec. 27, 1899), DASL 1:44.
26. "Hanlinyuan shihjiang xueshi Zhu Zumou zhe" (Memorial of Zhu Zumou, Expositor of the National Academy), Guangxu 25/11/24 (Dec. 26, 1899), DASL 1:43.
27. "Junjichu ji shuli Shandong xunfu Yuan Shikai dianzhi" (Telegraphed Order to Shandong Governor Yuan Shikai from the Grand Council), Guangxu 25/11/27 (Dec. 29, 1899), DASL 1:46.
28. "Zhili zongdu Yulu zhi zongligeguoshi yamen dianbao" (Telegram of Zhili Governor Yulu to the Zongli Yamen), Guangxu 26/4/25 (May 23, 1900), DASL 1:101.
29. "Telegraphed Order to Governor Yuan Shikai of Shandong from the Grand Council," Guangxu 25/11/27 (Dec. 29, 1899), DASL 1:46.
30. "Junjichu jege sheng dufu shangyu" (Grand Council Edict to All Provincial Governors), Guangxu 25/10/19 (Nov. 21, 1899, DASL 1:73.
31. Cai E, "Gengxin jishi" (Accounts of 1900-1901), YHT 1:305.
32. "Shangyu" (Imperial Edict), Guangxu 25/12/11 (Jan. 11, 1900), DASL 1:56.
33. "Xinbu shangshu jian shutian fuyin Zhao Shuqiao deng zhe" (Memorials from the Minister of Punishments and Prefect of the Metropolitian District, Zhao Shuqiao, and others), Guangxu 26/5/3 (May 30, 1900), DASL 1:110.
34. "Zhanshi fu sijin ju xianma Tanji zhe" (Memorial from the Librarian in the

Supervisorate of Imperial Instruction, Tanji), Guangxu 26/5/16 (June 12, 1900), DASL 1:131.

35. "Junjichu ji xiebandashi Gangyi deng shangyu" (Imperial Edict to Assistant Grand Secretary Gangyi and others), Guangxu 26/5/20 (June 16, 1900), DASL 1:145.

36. Hu Sijing, "Lubeiji," YHT 2:503.

37. "Junjichu ji gesheng dufu deng dianzhi" (Telegraphed Instructions to All Governors from the Grand Council), Guangxu 26/5/30 (June 26, 1900), SLCB 1:187.

38. "Yushi Liu Jiamo zhe" (Memorial of Censor Liu Jiamo), Guangxu 26/5/27 (June 23, 1900), in DASL 1:178.

39. Hong Shoushan, "Shishi zhilue" (A Summary of Current Events), YHT 1:91.

40. SDAJ 1:390.

42. SDAJ 1:47.

43. Wu Yong, "Engzi xishou congtan," YHT 3:373.

44. Zhong Fangshi, "Gengzi jishi" (Accounts of 1900), dated 6/16 (June 26, 1900).

45. Yang Mushi, "Zhi yinwuchuzhang Yuqu guanchashu" (Report to Battalion Affairs Officer, Yuqu), YHT 4:351, 353.

46. Li Di, "Quanfei huoguo ji," pp. 207-208.

47. SDAJ 2:711.

48. SDAJ 1:127.

49. SDAJ 1:381.

50. "Shangyu," Guangxu 26/8/14 (Sept. 7, 1900), YHT 4:52.

51. "Shangyu," Guangxu 26/5/25 (June 21, 1900, YHT 4:162.

52. "Quanluan jiwen" (Reports of the Boxer Riots), YHT 1:175, 178, 179.

53. Li Di, "Quanfei huoguo ji," p. 207.

54. "Quanluan jiwen," YHT 1:194.

55. Ibid., p. 193.

56. Huibao, no. 380.

57. Guo Dongchen, "Yihetuan zhi yuangu" (Causes of the Boxers) in Shandong Yihetuan diaocha ziliao xuanbian (Selected Survey Materials on the Boxers in Shandong) (Jinan: Qilu shushe, 1980), p. 331.

58. Weixian zhi (Gazetteer of Wei County), juan 20.

59. "Baxian dangan, Yihetuan zhuanjuan" (Special File of Documents on the Boxers from Ba District).

60. Wang Zengqi, "Liaoyuan shicun" (Poems from the Liao Garden).

61. "Baxian dangan."

62. Guangyi congbao (The Masses) 8 and 10 (1907).

63. SDAJ 1:227.

64. "Quanluan jiwen," YHT 1:196, 212.

65. Sun Zhongshan xuanji (Selected Works of Sun Yat-sen), 1:174-75.

66. Lu Xun "Qingmo zhi qianze xiaoshuo" (Critical Fiction from the late Qing Period), in Zhongguo xiaoshuo shilue (A Short History of Chinese Fiction), p. 298.

67. Liening quanji (Complete Works of Lenin), 15:159.

LIN HUAGUO

Some Questions Concerning the High Tide of the Boxer Movement*

The high tide of the Boxer movement lasted only a short time, about two months. During this stage the movement underwent many twists and turns, leaving behind many questions requiring further research. In this essay, I have selected three of the most important issues for discussion.

How were the Boxers able to sweep into Beijing and Tianjin in June 1900?

The traditional explanation has held that the power of the Boxers had developed to the degree that the Qing government lacked the power or did not dare to use its armed might to stop them from entering Beijing and Tianjin. In the fourth volume of *Zhongguo shigao* (Draft History of China) [1962], Guo Moruo states this view: "By May of 1900 when the Boxers had advanced to the outskirts of Beijing and Tianjin, the revolutionary struggle of the masses came to threaten the headquarters of the Qing regime and the dynasty faced an immediate danger of overthrow." Hu Sheng wrote in "Yihetuan de xingqi he shibai" (The Rise and Fall of the Boxers) [1979] that after the Boxers' armed struggle was approaching Beijing, "The Empress Dowager Cixi understood it was dangerous to have mobilized the troops in the capital and so sent Gangyi to the Zhuozhou region [about sixty kilometers from Beijing] to 'persuade' or 'instruct' the Boxers, but just at that point the Boxers swept into Beijing."[1] Jin Jiarui states in *Yihetuan yundong* (The Boxer Movement) [1957]: "Along a line extending

---

*From Lu Yao, ed., Yihetuan yundong (The Boxer movement) (Chengdu: Pashu shushe, 1985), pp. 503-18. Translated by David D. Buck.

from Baoding to the Marco Polo Bridge, the Boxers had defeated the Qing military's 'pacification' plan and so the Qing
units were forced to take up defensive positions at several
critical locations. . . . The Qing military lost the initiative and
the main units of the Boxers were able to gain entrance into
the Beijing and Tianjin region."[2] He even describes their
entrance as a "liberation." These explanations are dissimilar,
but they have common elements: first, the Boxers occupied the
cities as a result of struggle between themselves and the Qing;
and second, none of these accounts takes into consideration the
influence of the foreign Powers' armed intervention. Consequently, none provides a thorough and complete explanation.

Certainly it is true that by May 1900 the Boxers' power was
already developed enough so that as they approached Beijing
they became a challenge to the dynasty. Still, to assert that the
Boxers' strength was capable of overthrowing the Qing or that
the Qing lacked the military might to stop the Boxers from
"liberating" Beijing is to exaggerate the threat they posed.

The Boxers' special qualities were their mass character and
lax organization. Although it would have been difficult for the
Qing forces to suppress them across the vast North China plain,
it was not hard for the Qing's main military forces to gather
for the protection of Beijing and Tianjin. In fact the Boxers'
capacity to carry out such an assault was far inferior to that of
the Taiping Northern Expedition. Although the Qing efforts to
suppress the Boxers militarily had met with setbacks on several
occasions, those losses were not serious, and the main forces of
the Qing never were committed.

How, then, were the Boxers able to enter Beijing and Tianjin in the early and middle part of June 1900? If we hope to
clarify this issue, we must see it not simply as a contest between the Boxers and the Qing dynasty, but must add a third
factor--the armed intervention of imperialism, which produced
a great alarm in the highest circles of the Qing government.

The immediate goal of the foreign military expedition was
the suppression of the Boxers. On this point they were not in
conflict with the Qing, but the problem lay in another possibility. After the Sino-Japanese War of 1894-95 there was
constant talk of the foreign Powers occupying harbors and

carving out "spheres of interest." With the Allied armies poised, ready to proceed into Beijing and suppress the Boxers, could they not overthrow the Qing dynasty and proceed with the break-up of China? This is what gave the Qing government cause for concern. They were caught on the horns of a dilemma, for following the failure of their efforts to defeat the Boxers, the Boxer challenge to the dynasty was combined with a serious threat of a foreign invasion of the capital city.

How could the Qing government extract itself from the dilemma? Different groups at the Court proposed various solutions. The strongly xenophobic conservatives advocated halting the suppression of the Boxers and instead employing the Boxers' strength to oppose the foreign threat. On May 30, President of the Board of Punishments Zhao Shuqiao and Prefect of the Metropolitan Prefecture He Naiying jointly submitted a secret memorial which argued: "The Boxers are growing rapidly and so it would be better to employ them after placing them under military command and discipline than to continue to try to suppress them. Their hatred for Christianity could be made into the stuff of valor, and thus their personal grievances could be made to serve the public good. They would become more dependable and their power could thus be controlled."[3] Zhang Zhidong, Sheng Xuanhuai, and others in the Westernization group advocated the firm suppression of the Boxers, believing that only such a policy could forestall a foreign armed intervention. Governor General Yulu, who was then in command of military affairs in Zhili, also advocated a policy of firm suppression. He advised: "The followers of these outlaws [the Boxers] increase every day. They cannot be made peaceful by means of decrees or speeches. If we do not use military force to keep them under control, then I fear they will spread like a prairie fire."[4] Yulu went on to advocate the transfer of fresh troops to carry out a "three-pronged encirclement."

The Empress Dowager did not favor suppressing the Boxers. She was even more reluctant to use the Boxers against the foreign Powers, but with danger on all sides she could not afford to vacillate, and so she followed Yulu's suggestions. Just at this moment she received a memorial reporting that the Powers were demanding to send their troops into Beijing, and so her announcement of stringent measures of suppression against the Boxers was an attempt to appease foreign imperialism.

At this juncture, it was hopeless to have thought the Boxers could be speedily suppressed by force. The Powers took no notice of the Qing government's conciliatory position and dispatched more troops toward Tianjin and Beijing. The Empress Dowager was forced to find another approach: she adopted a new policy of working with the Boxers to end their violence. On June 5, Zhao Shuqiao was sent to the Zhuozhou region to exhort the Boxers to disband. The next day Grand Councillor Gangyi was dispatched to join him in this task. On that same day, June 6, the Court publicly decreed: "Christians and Boxers (*quanmin*) both are children of our state, and the Court cares equally for both of them."[5]

We must note that the Empress Dowager's views remained different at this time from Zhao Shuqiao's. She was not considering the use of the Boxers to oppose foreign aggression, but rather was trying to use peaceful methods to disband the Boxers. She hoped to calm the Boxers' anti-Christian activities and thereby take away the excuse the foreigners had for armed intervention.

When the Empress Dowager announced that she "cared equally for both of them" on June 6, she went on to forbid the Qing army from "injuring the people, or taking credit for the patriotic sacrifices of others." This language provided the basis for a temporary cessation of the Qing armed suppression against the Boxers and meant that any efforts to pacify them would have to be conducted with restraint. Grand Councillor Gangyi appeared to be carrying out the Empress Dowager's "explicit instructions to disband" the Boxers at Zhuozhou, but he tacitly had agreed that the Boxers should be tolerated. He was thus able to continue his plan to enlist the Boxers against the foreigners. Upon arriving there Gangyi used his own authority to order the Qing troops to halt their suppression of the Boxers and further directed the withdrawal of those Qing troops who had been fighting with the Boxers.

This sudden change of policy by the Empress Dowager, combined with Gangyi's forceful action to stop the suppression of the Boxers, left the local officials in Zhili who had been conducting the suppression in a quandary. They temporarily discontinued their military operations, but on June 9 Zhili Finance Commissioner Ting Jie and Chief Provincial Judge Ting Yong sent a telegram to Governor General Yulu sounding an alarm: "We have withdrawn to the north, but in the south of the province the Boxers' power grows ever stronger."[6] Their words provide a snapshot of the conditions that existed in the province at that moment. The Boxers took advantage of this situation to rush into Beijing while the Qing military machinery had fallen into semiparalysis.

The Boxers moved into Tianjin slightly later than they entered Beijing. One important reason for this was that among the nobility and high officials responsible for the defense of the capital were many, such as the Manchu nobles Zaiyi, Zailan, Zailian, and Zaixun,[a] who favored the Boxers. These people took advantage of the Empress Dowager's change of policy to let the Boxers enter the capital. However, at Tianjin, Yulu and Nie Shicheng were in command, and they firmly advocated suppressing the Boxers, so the impediments to the Boxers were greater there. The Boxers moved into Tianjin only after Admiral Seymour's forces had started out for Beijing, thereby sharpening the contradictions between the Qing Court and the foreign aggressors.

The anonymous author of *Tianjin yiyueji* (Record of a Month at Tianjin) reveals quite clearly the conditions that prevailed there at that time. As that account states, Boxer activity in Tianjin always had been strictly forbidden, and only on June 13 did a change occur. On that day at the Sanyi Temple in the city "a group of altars were set up," but the authorities did not come to disperse them. These altars (*tan*) were the organizational units used by the Boxers, and thus "both inside and outside the city the news that the altars at the Sanyi Temple would not be prohibited led to the setting up of many others."[7] This is how the Boxers organized so rapidly in Tianjin.

The general situation in early June 1900 had the Qing dynasty under threat not only of the Boxers but also of a foreign invasion. This dilemma forced the Qing to change its policy from one of strict suppression of the Boxers. The Boxers took advantage of this opportunity to move large numbers of their supporters into Beijing and Tianjin. This situation was the product of the complex three-way struggle involving the Boxers, the Qing government, and the foreign Powers.

Why did the Qing dynasty declare war on the foreign Powers?

Two explanations are the most common. First, Fan Wenlan in his book *Zhongguo jindaishi* (Modern Chinese History) [1953] states: "During the night of June 16 Grain Intendant Luo Jiajie sent a secret report to Ronglu consisting of four items, one of which stated that the foreigners wanted 'to compel the Empress Dowager to restore the Emperor to his authority.' The next day at down, Ronglu reported this to the Empress Dowager. She became furious and without further inquiry into the situation decided to risk everything in one throw of the dice."

Second, *Zhongguo shigangyao* (An Outline of Chinese History) [1964], under the general editorship of Jian Bozan, in its modern history section edited by Shao Xunzheng does not mention this report from Luo Jiajie. Rather, Shao Xunzheng emphasizes the threat to the dynasty posed by the Boxers anti-imperialist struggle and concludes that the dynasty was forced to declare war against the Powers because of this enormous pressure from the Boxers. Jin Jiarui's *The Boxer Movement* [1957], Guo Moruo's *Draft History of China* [1962], and Hu Sheng's "The Rise and Fall of the Boxers" [1979] all discuss the pressure brought to bear on the Qing by the Boxers, but they also mention Luo Jiajie's report. Yet in the end they follow the second line of interpretation, emphasizing the ability of the Boxers to pressure the Qing into declaring war. According to this second viewpoint, the Empress Dowager did not decide to act rashly on the basis of groundless rumors, but rather took a carefully considered decision in which the declaration of war became of a means for using the foreign armies to destroy the Boxers and thereby dispel their threat to the dynasty.

Fan Wenlan's main evidence is *Congling chuanxinlu* (A Collection of Letters from the Guangxu Reign [1875-1908]); Shao Xuncheng's evidence for the other interpretation is drawn from a Court telegram to the provincial governors on June 26 and a second telegram of June 29 to the foreign ministers residing in Beijing. The author of the letters, Yun Yuding, had participated in the imperial conference during which the Qing Court discussed the question of declaring war; consequently his explanation carries considerable weight. The Court telegrams explaining the reasons for going to war also are important evidence on this matter. Both viewpoints seem to have some basis, but their authors only use that evidence that supports their own interpretations and ignore what does not. Consequently, we have an overly simplistic analysis of this complex issue.

To clarify how the Empress Dowager came to declare war we must begin in the early days of June 1900. As discussed in the first part of this essay, the foreign military approach toward Beijing and Tianjin posed a serious threat to Qing rule, and an element among the conservatives advocated, beginning in early June, the use of the Boxers to resist this foreign advance. But the Empress Dowager was attempting to disband the Boxers and thereby obviate the foreigners' reasons for bringing troops into the capital city.

As the situation developed, the Empress Dowager's plans were upset. Not only did the Boxers remain unpacified, but they took advantage of her policy of toleration to enter Beijing and become more active. On June 8, she ordered that all Boxer activity in Beijing be stopped, but the Manchu princes and high officials implementing this order were mostly the same conservatives who favored using the Boxers and therefore did not carry out her wishes. Even more important is the fact that many of the Qing army units stationed in Beijing had joined the Boxers or become participants in their anti-imperialist activities. The Gansu Army under the command of Dong Fuxiang displayed this tendency most clearly.

As early as June 13 the Empress Dowager had decided that if the foreign troops continued their advance, she would resist by fighting. Yet, she strove to keep her resolve secret and, while preparing to go to war, adopted the pretense of continuing to suppress the Boxers and acquiescing to foreign demands. That very day in an imperial decree she renewed the defamation of the Boxers as "outlaws" and despaired that "the hope of securing the state might rest with the actions of these disorderly people." She went on to command that the Boxers be "strictly proscribed" and "arrested and gotten rid of" in Beijing. Further, she ordered Ma Yukun, stationed at Shanhaiguan in northern Zhili, immediately "to lead his troops to Beijing and en route carry out strict suppressions."

On June 15 she ordered Li Hongzhang "to proceed to Beijing with all dispatch" and Yuan Shikai "to plan to bring troop units along when coming to Beijing." Both these officials were subservient to the foreigners and known to be anti-Boxer. Asking them to come to Beijing obviously increased the pressure on the Boxers and was meant to please the foreigners. This proves that at that time the Empress Dowager feared foreign aggression more than she feared the Boxers. Her efforts to prepare for war by assembling regular troops in Beijing and to continue suppressing the Boxers were intended to halt the advance of the foreign armies toward Beijing. Thus, it was not pressure from the Boxers that shaped the Empress Dowager's policies, but rather the threat posed by the foreign Powers.

On June 15, Zhili Governor General Yulu formally reported to the throne, "the Powers have already dispatched troops and continue to reinforce their soldiers." He voiced his opposition to Cixi's order to use armed resistance in hopes of stopping the foreign troops' advance toward Beijing when he wrote: "Our power is completely inadequate in the face of the allied enemy troops." He added: "I do not dare to order [our troops] into combat, for even if I were to lead the preparations for battle myself, I doubt the troops could be made ready to fight."

Because the foreign troops continued their advance while Yulu dared not resist, the Empress Dowager concluded that the situation was even more critical. Yet the Boxers who had been ridiculed and cursed by the authorities as "rioters" were able to marshall their fearless spirit to resist the enemy. This made the Empress Dowager conclude that she must reconsider and find a more appropriate policy.

On June 16, she called an imperial conference. After the meeting a decree was issued ordering that the authorities should "leave the Boxers a way out." This marks the second time the

Empress Dowager ordered a halt to efforts to suppress them. At the same time, Grand Councillor Gangyi and General Dong Fuxiang were ordered "to enroll as troops the young and strong ones." Obviously the Empress Dowager had changed and was considering seriously the conservatives' suggestion that the Boxers be enlisted against the foreigners. The same day she ordered Governor General Yulu and generals Nie Shicheng and Luo Rongguang to block the foreign army's route of advance toward Beijing. She further ordered that if the foreign advance continued, "You high officials, without any thought of your own safety, must manage according to the situation confronting you, for the Court is too far away to control affairs. No responsible official can be permitted to hold back and endanger the overall situation." This gave Yulu the power to make war if the foreign army continued its advance. On the same day a similar order was given to Zeng Qi, commander of the Shengjing Army: "If the foreigners use the railroad to move troops toward Beijing, then you must find a way to stop them. If your troops are reluctant, you must lead them yourself. You must manage according to the situation confronting you, for the Court is too far away to control affairs."

These decrees reveal that even before Luo Jiajie had submitted his secret report on June 16, the Empress Dowager had decided about declaring war. If the foreign powers continued to advance their troops then war would ensue. Yet, the Empress Dowager was still fearful about declaring war on the foreigners. She continued to the very last to try to avoid a war.

On June 16, in response to a foreign demand, she sent the Imperial Guards to bolster the safety of the foreign legations, hoping that this might be sufficient to produce a halt in the foreign army's advance. That hope was destroyed quickly. In June, Yulu reported to the throne that the foreign commanders had issued an ultimatum that if the Qing troops did not hand over the Dagu forts, they would capture them by force.

At that moment the Empress Dowager realized the situation probably was lost. On June 19 she gave Yulu an order that read: "The situation has become critical. The soldiers have already begun to sacrifice themselves [in battle], and the high officials must quickly assemble brave men and resolutely lead the people to assist the army at every step in their resistance. There must be no cowardly withdrawals that would permit the foreign troops to advance. If the Dagu forts are lost, these same high officials will answer for it." This was not a formal declaration of war, but certainly it was an order to her own commanders to begin a war.

As to the question of why the Qing Court did not make a formal declaration of war on June 19, I believe the only reason was they did not know if the foreigners actually would carry out their threats against the Dagu forts. On June 20, the Empress Dowager told Yulu to inform the Court "if the battle to seize the Dagu forts has commenced." Yulu replied the same day that the aggressors' forces already had landed at Dagu and Tianjin, while the Qing army had joined forces with the Boxers to counterattack. After receiving this memorial, the Empress Dowager made her formal declaration of war the next day on June 21. She included an order for the Qing forces and the Boxers "to join as one" in resisting the foreign invasion.

From this account we can see that the decision to declare war did not come about suddenly. Only after Admiral Seymour's forces began their advance toward Beijing did the Empress Dowager come to a decision that it would be necessary to go to war. After the imperial conference on June 16 this decision was reaffirmed. The fact that the declaration of war was made *after* the taking of the Dagu forts proves the correctness of my version. It also shows that the declaration of war was not a rash decision made during a crisis. Fan Wenlan's version that after the Empress Dowager received the secret report of foreign demands for her to relinquish power she became "dizzy with fright" and decided "to risk everything on one throw of the dice" simply does not square with the facts. Fan Wenlan's evidence all comes from Yun Yuding's account. But careful reading of Yun's account show that the two matters of a declaration of war and the foreign demand for the Empress Dowager to give up power were not part of the same discussions. Linking them together is an interpretation placed on events by Yun Yuding and other like-minded officials but has no basis in fact.

This can be shown because there are some contradictions in

Yun Yuding's account. On the one hand, he recalled that after hearing Luo Jiajie's secret report, the Empress Dowager's "concern turned to anger and she decided to go to war." Yet he also records that *prior* to seeing Yun's report she had taken the position in an imperial conference that General Dong Fuxiang should enlist the Boxers to resist the foreigners. In addition, Yun Yuding recalled: "A certain official [Luo Jiajie] had taken on trust the words of others, but the foreign representatives had never made such remarks [about Cixi giving up power]. Therefore no declaration of war was issued until the 25th day [June 21]."[8] This means the Qing Court knew Yun's report to have been false prior to declaring war on June 21 and so omitted mentioning this matter in the text of the declaration. Since this is the case, it is impossible that this same report could have been the principal cause of the Qing declaration of war.

Fan Wenlan's other source is the "Jingshan riji" (Diary of Jingshan). I will not take up here the question of its authenticity, but there is a detailed account by Cheng Mingzhou in *Yanjing xuebao* (Yanjing University Bulletin) 27 (1940) asserting this diary is a forgery.[b] At the very least we can show that the narrative account in this diary about the declaration of war cannot be believed. The diary states: "On the 24th (June 20), Prince Duan [Zaiyi] informed the Grand Council and other high officials to assemble for an imperial audience at which the Old Buddha would be asked to surrender her control over the government."[9] This account is obviously false, and thus the diary cannot be used as the basis for any account of these events.

From this discussion we can see that for the Empress Dowager the greatest danger came from the foreign invasion and not from the Boxers. Her policy toward the Boxers was subordinate to her foreign policy. In early June her attempts to pacify the Boxers had been made in hope of heading off a conflict between the Boxers and the foreigners that would lead to a major foreign intervention. Her orders after June 13 rigorously to suppress the Boxers were made to please the foreigners and to forestall them from sending troops to Beijing. Her second order to stop the suppression of the Boxers, made on June 16, was given following the failure of her attempts to appease the foreigners. Only then did she consider enlisting the Boxers to resist the foreigners. After the foreign armies attacked the Dagu forts, she ordered "linking up" with the Boxers to launch a counterattack. Thus we can see that there never really was any pressure from the Boxers on the Qing dynasty to declare war, but rather it was the foreign invasion that forced the Qing into a war and caused them to employ the Boxers as part of their resistance.

**Qing policy toward the Boxers after
the declaration of war**

This issue is connected closely with the previous topic, for according to Fan Wenlan, the Empress Dowager's declaration of war was made when she was "dizzy with fright," and then after a few days she "gradually regained her senses." Thereupon on June 25 she "decreed a halt to the siege of the legations" and "reissued a strict suppression order" against the Boxers.[10] According to Shao Xunzheng, "The Empress Dowager's purpose in declaring war was not to fight the foreign powers, but rather to find a means, at last, of destroying the Boxers." Thus the policy of the Qing after declaring war became "apparently to be at war with the foreign Powers, but actually to have capitulated to them. At the same time falsely maintaining a front of cooperation with the Boxers, but actually awaiting an opportunity to destroy them."[11] The common threads in these two interpretations are that both argue there was no true cooperation between the dynasty and the Boxers against the foreign Powers, and both agree that the Qing policy fundamentally was opposed to the Boxers. I do not believe that these explanations coincide with the facts.

In the previous section I have already described how the Empress Dowager seriously considered using the Boxers to oppose the foreigners on June 16. Then on June 20 Yulu reported that the invading forces had taken Dagu and Tianjin, but the Boxers and the Qing troops were cooperating and "fighting victoriously" in defense. When the Empress Dowager declared war on June 21, she also gave instructions to encourage the

Boxers and went on to inform the chief provincial officials: "These are good people and their like can be found everywhere. If the governor generals and governors will assemble them into militia units and use them to resist foreign aggression then we will have success. The coastal and riverine provinces should act on this with special dispatch."

On June 26 Yulu was ordered "to assemble the Boxers" so as "to increase our military might" and recapture the Dagu forts. On June 28 Yuan Shikai was ordered "to assemble quickly the Boxer militia (Yihetuan *yong*), to supply them with arms and food" in order to reinforce the North. On July 1, an order was issued "to abolish all prohibitions" against the Boxers that might still be in force. During this period there were a great many such orders.

If the Qing did not really intend to enlist the Boxers against the foreign Powers, or if they intended to fool the Boxers in order to destroy them, why would orders be issued "to assemble" Boxers and "provide them with arms"? If the Qing Court really viewed the Beijing and Tianjin area Boxers as their most dangerous enemy, why would the Court have given Yuan Shikai such orders to reinforce the capital region with Boxers from Shandong? The Qing Court's orders were opposed by the capitulationist group, which included Li Hongzhang, Zhang Zhidong, Liu Kunyi, and Yuan Shikai, but in some locations the Court's order to support the Boxers was implemented. Thus, in Tianjin the Qing Army and the Boxers certainly conducted a joint resistance, and the Boxers grew stronger in the process. In Manchuria, Shanxi, Inner Mongolia, and Henan, the Boxers acquired status as a legal organization and thus could develop more rapidly. These all are incontrovertible facts.

It is true, however, that in the defense of Tianjin the troops of Nie Shicheng did attack the Boxers, but this in itself cannot be proof that the Qing policy toward the Boxers was "falsely maintaining a front of cooperation with the Boxers while actually waiting to destroy them," because General Nie Shicheng was not following Qing policy and lacked official approval for his actions. In fact, he later was reprimanded for his troops' clash with the Boxers.

Of course the contradictions between the Qing state and the

Boxers were fundamentally antagonistic, and thus the ultimate goal of the Qing was to destroy the Boxers. There could not have been any long-term cooperation between them. Still, that does not exclude the possibility that in the face of an imminent imperialist invasion, the Qing and the Boxers could have temporarily joined forces to counter a common enemy. Hu Sheng and others deny that the Qing and the Boxers cooperated in fighting the foreign army because they also deny that there existed any major contradictions between the Qing dynasty and the imperialists, and thus they deny that a large-scale invasion by the foreign Powers posed a major threat to the dynasty.

In his analysis of the foreign Powers' policies, Hu Sheng asserts: "In this war the aggressors clearly indicated from the beginning that they were going to crush the Boxers to provide an example of just what they could do to any Chinese who dared to oppose imperialism. They did not invade because they felt the Qing dynasty was their enemy. The truth of the matter was exactly the opposite: the foreign Powers intended to rescue this government in order to stop the Qing from continuing to be compelled to do unthinkable things while under the domination of riotous mobs."[12]

In analyzing the position of the Qing government, Hu Sheng believes: "The Qing already had become the docile and obedient slave of the foreign Powers, and then it suddenly fell captive to the Boxer anti-imperialist movement. The foreigners were determined to rescue the Qing from this situation by whatever means were necessary."[13] According to Hu Sheng's analysis, the important issue facing the Qing at this moment was how to free itself from Boxer domination, whereas the goal of imperialism in invading China was to "rescue" the Qing from this "domination." This seems to hold together, but if this really had been the case, then the Qing dynasty never could have rejected foreign demands and certainly never would have cooperated, even temporarily, with the Boxers in opposing their "rescuers." The facts do not support Hu Sheng's arguments.

We know that once the Allied forces occupied Beijing they did not demand the break-up of China, nor did they overturn the Qing dynasty. Rather they continued to use the Qing state as the recognized representative of China in diplomatic matters.

The Qing responded with deep gratitude, but that is not grounds to conclude that either before or during the Boxer incident the imperialists were not planning to carve up China. In general, the final goal of all imperialist aggression against more backward states is to turn them into colonies. The status of semi-colonialism is only a transitional status, and imperialism's basic nature cannot be satisfied with this variety of indirect control. Prior to the Boxer movement, imperialism had occupied China's port cities and carved out spheres of influence in preparation for turning the whole of China into its colonies. After the Allied army occupied Beijing, China was brought face to face with the threat of dismemberment.

This threat forced a change in the attitudes of the Chinese ruling class. Not only were the patriotic landlords and officials and conservatives of a purely xenophobic cast demanding resistance to the foreigners, but even the Empress Dowager, famous for her fawning and subservient posture toward the Powers, had to take part in the struggle to preserve the Qing dynasty.

Both the Empress Dowager and Li Hongzhang are infamous traitors in modern Chinese history. Their kind generally were compromising and made concessions, but these two individuals' attitudes were somewhat different when faced with a foreign threat to the very existence of the dynasty. For those people of Li Hongzhang's ilk, if China were free from occupation or dismemberment, they could wield great power through their positions as important Court officials and representatives trusted by the foreigners. If China were to be occupied or divided up by the Powers, these same individuals would continue to occupy positions of power by becoming colonial officials. They knew that in semicolonial China it was the imperialists and not the Qing who were in control. In the Boxer incident, they acted so as to lose neither the Qing dynasty's trust nor the imperialists' favor. Once open foreign armed aggression was underway, they did not resist but continued to cooperate with the foreigners within their jurisdictions and to suppress popular anti-imperialist struggles.

On the other hand, the Empress Dowager, who held the dominant power at the Qing Court, realized that foreign occupation and dismemberment would bring about the total collapse of her own power and that of the Qing dynasty which she represented. So she maintained a fixed opposition on the foreign invasion. The position of most of the nobles at the Qing Court was the same as that of the Empress Dowager. Consequently, in the controlling circles of the Court those advocating war prevailed for a short while. Faced with strong aggression, they temporarily put aside their differences with the Boxers and cooperated with them.

The corrupt Qing administration could not carry out this opposition to imperialist invasion in a thorough manner. Also a crafty old fox like the Empress Dowager was not willing to stake her whole political fate on a policy of war against the foreigners. In spite of this, in the declaration of war she had vehemently stated: "To drift along with them [the Powers] hoping to survive would be to shame our ancestors. If one has acquired a mighty weapon, it is best to seek a showdown." Yet, in truth, she did not dare "seek a showdown" and really only could hope "to drift along with them hoping to survive." The Empress Dowager was forced to resist, but even in resistance she sought compromise.

Proof that the Qing government sought to compromise with the foreigners can be found in the Court's policy in regard to the foreign legations in Peking. They appeared to advocate attacks on the legations, but really acted to protect them. By that time the Powers would not consider halting their advance and intended to widen the war. This forced the Qing to seek peace while still continuing to fight. The Qing government followed the battles at Tianjin with great interest and several times urged Yulu and the Boxers to join forces to take the foreign concession there, and then to recapture the Dagu forts. They also repeatedly urged the provincial governors to send troops to assist at Tianjin, while at the same time reprimanding Ronglu and Nie Shicheng for the reverses they suffered in battle.

After Tianjin had fallen, Beijing lay exposed before the invading army, and the Qing government became extremely fearful. As a result they increased their peace-seeking activities, but the commanders of the invading army refused to talk with the Qing and continued their advance. The Qing had no choice but to continue their resistance. The Empress Dowager ordered

Yulu and Song Qing to "try every means . . . and use all your strength to forestall defeat by checking the enemies' northward raid." She also urged the provinces to send troops to aid Beijing and charged Li Bingheng of the prowar party with the responsibility to lead battles in the front lines.

By this time peace-seeking already had become dominant, while resistance was conducted only to bolster the Qing position in any peace talks. Thus, because the Qing was continuing to resist, they had to maintain some degree of cooperation with the Boxers. To pave the way for eventual capitulation, the Court could not let the Boxers' power continue to expand. Thus while continuing to use the Boxers, the Court also increased its control over and damage to them, hoping slowly to weaken them.

The Court discontinued the calls to assemble Boxers in the provinces and stopped orders to the provinces to send Boxer reinforcements to Beijing. In addition they issued orders forbidding Boxers from the areas around Beijing to enter the walls of the city, while also transferring some Boxers from within the city and the suburbs to the front lines. These measures were taken in hopes that the Boxers and the foreign army would weaken each other in battle. In other locations the Qing renewed their efforts to stifle the struggle of the Boxers against the foreign invasion. Nonetheless, the cooperation between the Qing administration and the Boxers never was completely broken, especially along the front lines east of Beijing where the government forces continued to work with the Boxers to block the invading army's advance.

On August 14 when Beijing fell, and after the flight of the Qing Court, the situation underwent a fundamental change. The Empress Dowager realized that it was impossible to save the Qing dynasty by continuing the policy of resistance, so she elected to change to a policy of complete capitulation in hopes of obtaining "leniency" from the foreigners. Thus the Boxers had lost all value in her eyes and became only an impediment to peace. Her policy toward the Boxers had changed from a two-faced policy that combined cooperation and use in one aspect while also seeking to control and weaken the Boxers in the other. It was replaced by a straightforward policy of open suppression.

This account of the period after the formal declaration of war has emphasized the point of view of the Qing Court in its analysis of the relations between the Qing and the Boxers. Using this point of departure helps clarify the situation and its lessons.

Notes

1. Jindaishi yanjiu (Research on Modern History) 1 (1979).
2. Yihetuan yundong (The Boxer Movement) (Shanghai: Renmin chubanshe, 1957), pp. 56-61.
3. Guojia danganju Ming Qing danganguan (Ming-Qing Archive of the National Archive), eds., Yihetuan dangan shiliao (Archival Materials on the Boxers; hereafter DASL) (Shanghai: Zhonghua shuju, 1959), 1:110.
4. DASL 1:113.
5. DASL 1:118.
6. Yihetuan yundong shiliao congbian (Collected Materials on the Boxer Movement) (Beijing: Zhonghua shuju, 1964), part 2, p. 183.
7. Qian Bocan et al., eds., Yihetuan (The Boxers, hereafter YHT) (Beijing: Shenzhou guoguangshe, 1951), 2:141.
8. YHT 1:47-49.
9. YHT 1:66.
10. Zhongguo jindaishi (Beijing: Xinhua shudian, 1951), 1:360-61.
11. Zhongguo shigangyao 4:100-101.
12. Jindaishi yanjiu 1 (1979).
13. Ibid.

Translator's Notes

a. Three of these men, Ziyi, Zailan, and Zailian, were the sons of Prince Dun, Yizong, the younger brother of the Xianfeng emperor (r. 1851-1862). Zaiyi became the most prominent member of his family after his father's death in 1889. He inherited much of the property of an uncle and a title; in 1894 he was designated Prince Duan. His rise was due in part to his wife, Guixiang, who was a niece of the Empress Dowager Cixi.

Prince Duan was one of the first courtiers to support the Empress Dowager against the Guangxu emperor's reform program in 1898. After that crisis his power increased. In early 1900 his son, Pujun, became the officially designated heir to the throne. Prince Duan became head of the Zongli yamen in June 1900 and generally is believed to have been one of the chief advocates of using the Boxers against the foreigners. After the collapse of the Boxers, Zaiyi was ordered banished to Ili in 1901 and was to be imprisoned there for life, but the dynasty fell while he was still vigorous and he returned to Beijing after 1911. In 1901 his son was removed as heir apparent. Zailan and Zailian were less important but still numbered among the top pro-Boxer nobles at the Court in 1900. Both were punished for their roles in the Boxer movement.

Zaixun was another prominent Manchu nobleman of the same generation who inherited the title of Prince Zhuang in 1875. He opened his Beijing residence to the Boxers as a camp and office ground in 1900. Zaixun was ordered to commit suicide for his role in the siege of the legations and did so in February 1901. See Arthur Hummel, Eminent Chinese of the Ch'ing Period (Washington, D.C.: Government Printing Office, 1943-44), 1:393-94; 2:926.

b. The dispute over the authenticity of this diary has been a long and fascinating one. The diary was translated into English in 1924 by Professor J. J. L. Duyendak (Acta Orientalia, 3), who expressed some concern over parts of the diary but generally accepted it. William Lewisohn challenged its authenticity in an article, "Some Critical Notes on the So-called 'Diary of His Excellency Ching-shan.' " in Monumenta Serica 5 (1940): 419-27. See Victor Purcell, The Boxer Uprising: A Background Study (Cambridge: Cambridge University Press, 1953). For the story of the diary's finder and first translator see Hugh Trevor-Roper, The Hermit of Peking: The Hidden Life of Sir Edmund Backhouse (London: Penguin, 1978).

HU BIN

Contradictions and Conflicts Among the Imperialist Powers in China at the Time of the Boxer Movement*

The Boxer movement of 1900 was a patriotic movement of the Chinese people against imperialist aggression. It shook the world by dealing a heavy blow against the aggressive presence of the imperialist Powers in China. To drown it in blood, the Powers sent over 100,000 troops to China. During those days, the Boxers' suppression became the Powers' common objective, and they managed to achieve a superficial unity, although there were many contradictions among their ranks even before they had captured Beijing. With the fall of the capital, the complex contradictions and conflicts among the Powers became more apparent with each passing day. The present article reveals and analyzes how these contradictions and conflicts developed into a temporary compromise among the imperialist Powers.

I

The spring and summer of 1900 witnessed a considerable growth in the influence of the Boxers in the Beijing and Tianjin region. Panic-stricken, the diplomatic corps of the imperialist Powers at Beijing decided to summon their soldiers to the capital on the grounds of protecting their legations. On June 9 British Minister Claude MacDonald learned that the Boxers were approaching the suburbs of Beijing. Sensing the tension in the capital, he telegraphed the British consul at Tianjin, ordering him to instruct Admiral Seymour to dispatch

*From Yihetuan yundong shi taolun wenji (Collected Articles on the History of the Boxer movement) (Jinan: Qilu shushe, 1982), pp. 338-57. Translated by Zhang Xinwei.

strong reinforcements to Beijing without delay. On the following day, Seymour mustered an international force in Tianjin
and started an advance toward the capital. On the way, his
forces were intercepted and held at bay by Qing troops and
civilians. Only after two weeks of bitter fighting did Seymour's
column manage to return to Tianjin.

The failure of Seymour's expedition was caused mainly by
the valiant resistance of both the Boxers and the Qing troops,
but the contradictions among the Powers also played a role.
When Seymour was gathering his invasion force, "the French
and the Russians refused to dispatch guards, unless a force of
at least 1,500 men was sent or the [railroad] line was in working order."[1] Yet Seymour left Tianjin in a great hurry, with no
more than 600 to 700 men. Only later, with the arrival of
French and Russian reinforcements, did his column's strength
reach 2,000. Its failure gave rise to increased mutual suspicions
among the Western Powers. German Foreign Secretary Prince
von Bulow said: "The failure of the expedition is attributed less
to military obstruction than to political opinions of the commanders. The intense hatred and suspicion existing between the
Anglo-Japanese bloc and the Russo-French bloc makes it most
unlikely for the Powers to keep their unity to the end of the
dispute."[2] Except for underestimating the role of the Boxers,
von Bulow's analysis of the contradictions among the Powers
tallied with the actual situation.

The failure of the Seymour expedition, the murder of German Minister at Beijing Baron von Ketteler, and the siege of
the legations in Beijing were all used by the Powers as excuses
for an expansion of their aggression. Sending reinforcements to
China one after another, they felt the need to organize an international force to coordinate the operation.

The position of commander-in-chief for this Allied force
became a bone of contention. Some held that this international
force should be under the control of the senior commander
present, while others thought that the number of troops contributed by each nation should determine the selection. Neither
position met with the approval of Russia and Britain, the two
Powers with the greatest presence in China. The Russians
pointed out that appointment of the commander-in-chief on

the basis of either seniority or the numbers of troops, "being subject to frequent changes, is open to serious practical inconveniences."[3] Britain also maintained that "the appointment of a General Commanding in the pursuance of these considerations would not be satisfactory."[4] Yet, neither could propose a solution acceptable to the other Powers.

As a result, the contention among the Powers over the commander-in-chief's post became more intense. Russian Minister of War Kuropatkin said: "The condition of our participation in an advance on Beijing is that the Russian troops must not be placed under the orders of a British, Japanese, or American commander."[5] He advocated that "the general leadership of the troops should be given to Admiral Alexeieff."[6] Meanwhile, Britain indicated that "she is unwilling to see her troops under a foreign commander but will only act in concert with Japan."[7] Germany tried to use the differences between Britain and Russia to its own advantage. Kaiser William II believed that the failure of the Allied forces under Seymour "made Europeans lose face before the Asians." He said, "A British commander-in-chief is absolutely unacceptable to us." Reluctant to express his opposition publicly, however, he expressed his "hopes that the Russians and the French could spare us the disgust of refusal."[8] Thus, he tried to push France and Russia to the forefront in opposing Britain to deepen the contradictions between Britain and Russia.

When the murder of Baron von Ketteler was confirmed, the kaiser was outraged and decided to send more naval and land forces to China. At the same time, he began working to have a German appointed as commander-in-chief of the Allied forces. Early in August the German government proposed to the other Powers that Field Marshall Count von Waldersee be appointed to that position. The proposal was approved by Britain, Russia, and Japan in rapid succession; only France was slow to reply. As a result of the humiliation it had suffered in the Franco-Prussian War, France was unwilling to place its troops under German command. French Foreign Minister Declasse said: "France cannot agree to the appointment of a German general as commander-in-chief. She will not tolerate it. What is more, it will give rise to an imminent domestic crisis."[9] But as Waldersee's position as commander-in-chief of the Allied forces had already been recognized by the other Powers, France would find it difficult to exclude itself from their cooperation. In mid-August, France replied to Germany: "As soon as Marshall von Waldersee shall have arrived in China, and shall have taken in the Councils of the Commanders of the International Corps of the Army, the eminent position due to his superior rank, General Voyron, the Commander of the French Expeditionary Corps, will not fail to place his relations with the Marshall on a proper footing."[10] The reply was ambiguous. Though not a flat refusal, it was by no means a complete acceptance. The German government was satisfied, however, and ordered Waldersee to leave for China to assume his duties without delay. Thus, the rivalry among the Powers for the position of commander-in-chief of the Allied forces came to an end.

II

While sending troops to suppress the Boxers, the imperialist Powers tried to take the opportunity to grab more of China's territory, and they also hoped to consolidate and expand their spheres of interest in China. In the summer and fall of 1900 the crisis of China's possible partition reached an apogee.

In the late nineteenth century, Russia had begun to consider Northeast China as its exclusive sphere of interest. After the rise of the Boxer movement in the Beijing-Tianjin region, Russia plotted to occupy Northeast China by force and to turn it into a colony under its direct control. A Russian wrote: "It is an open secret that, from the very beginning of the campaign, it was the desire of the military party not only to punish the Boxers but also permanently to annex Manchuria."[11] Late in July 1900 Russia invaded in force the provinces of Heilongjiang and Jilin from the Siberian maritime provinces, while also advancing northward from the Liaodong peninsula. On August 8, the Russians occupied the southern Manchurian treaty port of Niuzhuang, hoisted their national flag over the Customs House, and established their administrative offices there. The same was done at Harbin in the North. By late October the whole of Northeast China had fallen under the control of the invaders

At the time of the Boxer movement, Britain was engaged in an aggressive war to conquer the Boxers in South Africa. Except for a number of warships and some army troops from India, it was unable to send much support to China. So, it tried to persuade Japan to take advantage of that nation's geographical proximity to China in order to dispatch rapidly tens of thousands of troops, "not only to save the legations and foreign subjects, but with a view to the repression of the insurrectionary movement provoked by the Boxers and the reestablishment of order at Beijing and Tianjin."[12] Britain even promised to provide Japan with aid of a million pounds.

It was Britain's hope to use Japan to oppose Russia's influence, leaving Britain to concentrate its main forces in China in its own sphere of influence--the Yangzi Valley. Late in July, when Russia was sending troops to invade Northeast China, Britain started a secret plot to occupy Shanghai and for that purpose consulted with Liu Kunyi, viceroy of Lianjiang. Arriving in Shanghai from Dagu, Admiral Seymour took an "inspection" tour of the foreign concessions and the suburbs. After that, he suggested to the British government that it should send 3-5,000 new troops to Shanghai. On August 9, he and C. W. Campbell, British consul at Shanghai, informed the consular body that the British forces would arrive there on the 12th. The consuls telegraphed their respective government, suggesting that the British troops, "should be landed in Shanghai under the international agreement. It is advisable that additional troops be immediately dispatched by the allies for the efficient protection of this port."[13] In other words, Shanghai should not be occupied by the British troops alone, but its occupation should be the concerted action of the Allied forces. On the 18th, 2,000 British troops landed in Shanghai, to be followed by additional men from Hong Kong; subsequently troops from France, Germany, and Japan followed. Britain's plan for an exclusive occupation was thus frustrated, and there emerged a new source of tension among the Powers that could have turned into an armed confrontation. It was not until 1902 that the Powers came to an agreement on the evacuation of troops from Shanghai. By that time their troops had garrisoned the foreign concession for three years.

In 1898 Japan extorted an agreement from the Qing government not to cede or lease Fujian province to other Powers and was attempting to make that southeastern coastal province part of its sphere of interest in China. After the Russian invasion of Northeast China and the landing of British troops in Shanghai, Japan became eager to join the game and plotted to occupy Xiamen (Amoy). On August 24, 1900, it landed troops there on the excuse that a Japanese temple had burned down. Upon learning of this, the British immediately ordered their warship *Goddess* to Xiamen with a detachment of marines to be put ashore there. Although the troops of the two nations were later removed, Gulangyu island opposite Xiamen was established as an international concession in 1901.[14]

While the other Powers were moving their troops about in an attempt to turn their spheres of interest in China into colonies, the American imperialists were afraid that their interests in China might be seriously damaged. Therefore, in a circular diplomatic note, Secretary of State John Hay reiterated the "Open Door Policy" of 1899, maintaining "the need to preserve Chinese territory and administrative integrity." The true intention of the United States, however, was only to demand that the Powers maintain the status quo in China and refrain from taking direct steps to partition China, so as to ensure that the U.S. interests there would not be infringed upon.

Shandong was the sphere of interest for German imperialism. It was located between the spheres of Britain and Russia. Britain wished Germany to expand its influence northward against Russia, whereas Russia would have liked to see Germany expand toward the Yangzi valley in order to deepen the contradictions between Britain and Germany. Concern for the security of the German homeland's eastern frontiers made Germany unwilling to offend Russia. At the same time, Berlin coveted the rich and densely populated Yangzi valley and hoped to elbow its way in to share with Britain the advantages in that region. Late in July, British Prime Minister Lord Salisbury indicated his hope that all military operations in the Yangzi territory should be reserved for Britain alone. Upon hearing this news the German government objected at once,

demanding joint protection for the concessions in Shanghai.
Then Germany advocated that the Allied forces should maintain a joint watch on Chinese warships on the Yangzi River.
Consequently, relations between the two countries became
strained for a time. Late in August, Edward the Prince of
Wales visited Germany. At a reception, Kaiser William II said
emphatically, "It is a politically essential principle for both
England and Germany that the Yangzi valley should be open to
all the Powers under equal conditions."[15]

Aware of its inability to garrison the Yangzi region alone,
Britain was compelled to concur with Germany and then entered into negotiations on this matter. To insure German interests in that area and "avoid as far as possible such interpretation of the agreement as directed against Russia,"[16] the
German draft only mentioned that the Yangzi valley should
remain free and open. The British responded with an amend-
ment extending the principle to all ports on the coasts and
rivers of China, including Manchuria. There ensued a fierce
wrangling between the two parties. The German ambassador in
London, Hatzfeldt, pointed out to Salisbury: "I do not see what
practical benefit England can get from this article. For one
thing, the trade in Manchuria is comparatively insignificant;
for another, we cannot really believe that the Russians will
allow themselves to be driven out of Manchuria easily. It is
quite a different story with the Yangzi valley, which is the
richest area in China and also has the most important share in
her trade with Europe."[17] In his reply, Salisbury denied that
Britain had any intention of infringing upon Russia's interests
and argued that the ports on the coast in Manchuria had long
been open to foreign trade, and that the dispute involved only
a few ports along the Amur River. He was prepared to make
an exception of them, he said, if Germany should disagree. On
October 16 the Anglo-German Agreement was signed after
over a month's negotiation. Its first article stipulates: "It is a
matter of joint and permanent international interest that the
ports on the rivers and littoral of China should remain free and
open to trade and to any legitimate form of economic activity
for the nationals of all countries without distinction; and the
two Governments agree on their part to uphold the same for all

Chinese territory as far as they can exercise influence."[18] Although the article was worded in accordance with Britain's intention, Germany declared that the agreement did not apply to Manchuria. Thus, in the form of a treaty, Britain and Germany reaffirmed the "Open Door Policy" of U.S. imperialism, which also had been approved by the other Powers. The "Open Door" mitigated for the time being the contradictions among the Powers in the area of China south of the Great Wall.

III

Instead of starting peace talks as soon as their forces arrived in Beijing on August 14, 1900, the imperialist Powers tried to postpone them by every possible means. On the 28th, British Minister MacDonald dispatched a telegram to Salisbury in which he said: "To delay negotiations would not, in my opinion, entail any loss upon us, seeing that it will not be possible for some time to come to arrive at a general settlement."[19] Meanwhile, as he was preparing to leave for China, Count von Waldersee, who had been appointed commander-in-chief of the Allied forces, clamored to "avenge" the murdered German Minister von Ketteler. William II put it even more brazenly: "In the near future, it will be force rather than diplomacy that will decide everything."[20]

Under such circumstances it was impossible to start the peace negotiations immediately, although the Qing government had already appointed Li Hongzhang and Prince Qing [Yi-kuang] as its plenipotentiaries.

During the fifty days following August 14, when the Allied forces entered Beijing, down to October 4, when France finally proposed the basis for peace negotiations, the Powers had differed greatly on two issues. First, should they withdraw their legations and occupation troops from Beijing; and second, should they accept Li Hongzhang as the plenipotentiary in the peace negotiations.

In regard to the question of the Powers' presence in Beijing, the Russian Foreign Office instructed its diplomatic envoys on August 25 to notify the governments of their host countries that the Russian government "intends to recall its minister of

the fourth rank, M. de Giers, and all the personnel of its
legation to Tianjin, and the Russian troops will escort them to
said place."[21] Obviously, Russia's note was only a gesture to
show its support for the quick return of the Qing government
to Beijing. With this proposal the Russians hoped to ingratiate
themselves with the Qing government, so that an agreement
concerning Northeast China could be concluded and Russia's
ambition for permanent control of that region could be
realized. As Russia's ally, France only could express support. Its
minister, Pichon, was instructed to leave Beijing with all the
staff of the legation and the French troops as soon as he consulted with the Russian minister and officers.[22] In fact, the
French did not leave Beijing with the Russians when the latter
withdrew to Tianjin.

The Russian proposal met with the opposition of all the
Powers except France. The Hong Kong Chamber of Commerce
telegraphed the British Foreign Office: "Hong Kong Commerce
Chamber earnestly advocates maintaining Allied force Beijing
until proper government established, and guilty officials
punished. Earlier withdrawal most disastrous foreign prestige
throughout China."[23] MacDonald in Beijing expressed similar
sentiments. Consequently, the British government gave a negative reply to the Russian circular.

Germany, Austria, and Italy chimed in with England. Germany was especially satirical, saying that Russia was acting like
a "lover paying court to China."[24] On receiving the Russian
circular, Japan declared that it "intends to recall that portion
which may be deemed technically superfluous, or which is in
excess of the number actually required." But the Japanese made
it clear that "in view of Japan's proximity to North China, it
would be relatively easy for her in case of need to send troops
again, and therefore, the Imperial Government feels assured
that no bad results will be caused by the measures they propose
to take."[25] As no government had accepted its proposal uncon-
ditionally, Russia was forced to relent. On October 9 Minister
de Giers was instructed to return to Beijing and to participate
in the preliminary talks among the Powers before the peace negotiations.

The Qing government hoped to rely on Li Hongzhang, viceroy of Liangguang, to negotiate the terms of capitulation on its behalf. On July 8 Li was transferred to be viceroy of Zhili and minister of trade for the northern ports. On August 7 the Qing government appointed him plenipotentiary for peace negotiations and urged him to proceed northward without delay. Although Li had arrived in Shanghai in mid-July, he chose to remain there waiting for the situation to develop while the Powers wrangled over his qualifications. Li Hongzhang was well-known as the leader of the pro-Russian faction in the Qing government after the Sino-Japanese War of 1894-95. He had also negotiated the Sino-Japanese Treaty of 1896. Consequently, Russia was more than willing to have him act as chief negotiator in the Boxer settlement. They considered his plenary powers to be sufficient and valid. Earlier, however, the commanders of the Allied fleets at Dagu had reached a joint decision: "If Li Hongzhang arrives at Dagu, he shall not be allowed to have any communications with the land." The admiral in command of the Russian fleet objected on the spot, but his objection was voted down. On August 17, Lambsdorff, Russian minister for foreign affairs, sent a circular to the Powers demanding cancellation of this decision. The Russian position met with the approval of the American government, which believed that Li Hongzhang should be accepted to represent the Qing dynasty in the peace negotiations.

Germany, however, remained adamant, accusing Russia of seeking at all costs--"a fast peace," "a rotten peace." Kaiser William II said: "We must create guarantees for our merchants and missionaries to perform their missions in safety. But when Beijing is in turmoil and the Chinese government--if it is still existing--cannot be reached, how should such guarantees be created now? So, it is premature to negotiate for peace, especially with Li Hongzhang."[27] He even contemplated capturing Li Hongzhang "as a valuable hostage" when the latter left Shanghai for the North.[28] Japan, which coveted Northeast China for itself, remained unwilling to let Russia monopolize that area and thus refused to accept Li Hongzhang as a representative for negotiations.

The British government was of two minds about Li's qualifications. On the one hand, it was apprehensive that he might throw himself into Russia's arms, thus causing an unfavorable change in the balance between Britain and Russia in China. On the other, it was afraid that if it did not recognize Li Hongzhang's plenary powers the delay might result in the collapse of the Qing dynasty, an outcome that was also against Britain's interest. On August 23, Salisbury notified the Chinese minister in London, Luo Fenglu, that the British government "could not take any decision" on Li Hongzhang's qualifications for the time being.[29] Apparently at that time the British were still hesitating about what position to take and preferred to take a wait-and-see attitude. Later, the British were forced to give way gradually. On September 27, Salisbury instructed MacDonald that "if the power of Prince Qing [Yikuang] and Li Hongzhang is deemed sufficient, and on clear understanding that negotiations are purely preliminaries for reference to Governments, you are authorized, in conjunction with your colleagues, to commerce negotiations with them."[30] Thus, although the British government still had some reservations, it accepted Li Hongzhang to negotiate the settlement.

On October 4, 1900, the French government dispatched a circular containing six points as a basis for negotiations with the Qing government. The six points were (1) punishment of the principle culprits, (2) prohibition of imports of arms, (3) indemnities for losses, (4) establishment of guard units at the legations, (5) dismantling of the forts at Dagu, and (6) foreign military occupation of locations on the road between Tianjin and Dagu in order to insure free communication between Beijing and the sea. The other Powers accepted the French proposal in principle but made amendments and additions. On the 16th, the French government declared in a revised circular: "All the Powers adhere to the principle of the French note. The points which have given rise to observations on the part of certain Cabinets could be discussed among the Powers, or between their ministers at Beijing, in the course of negotiations, and could be modified in such manner as might be considered necessary for the more speedy attainment of the common aim."[31]

In October and November the diplomatic corps in Beijing held a number of meetings to make additions and amendments to the French proposals, and these can be summarized as follows:

1. Additions to the list of principal culprits to be punished. At a meeting on October 26 the ministers agreed to demand that the Qing government execute those provincial officials who were responsible for the murder of foreigners, in addition to the princes and ministers directly responsible for the attack of the legations at Beijing. They placed special emphasis on the inclusion of Dong Fuxiang and Yuxian in the list of offenders.

2. The Qing government's responsibility for suppressing the antiforeign activities occurring in China. The American minister submitted this article at a meeting on October 31, and it was approved unanimously.

3. Revision of treaties of commerce and navigation, plus solution of outstanding commercial problems. The British minister Satow brought up this point at a meeting on November 5 but met with the opposition of the Russian Minister de Giers, who believed that "commercial questions did not come within the scope of the present negotiations." Satow replied: "It appears to me, therefore, that the Plenipotentiaries are entitled to include in the note a clause intended to promote the interests of commerce which were also unrelentingly assailed."[32] As neither side was ready to yield, the matter was put to a vote. The British proposal was carried with Austria, Belgium, Germany, Italy, and the United States in favor and Russia and France opposed.

4. The Italian minister, Salvago-Raggi, suggested adding another article that "China will take financial measures on the lines which will be indicated by the plenipotentiaries in order to guarantee the payment of the indemnities and the service of the loans."[33] The proposal was opposed by the ministers from Russia and France, but those of Austria, Belgium, Germany, Italy, and the United States indicated that they would recommend it to their respective governments for approval.

The six-point proposal of France eventually was enlarged to twelve articles through the diplomatic corps' meetings. The first secretaries of the Austrian, Italian, and French legations formed a committee to draft the "preface." Their document was presented to the Qing plenipotentiaries in the form of a joint diplomatic note, which later became known as the "Outline for Peace Negotiations." On December 24, Spanish Minister Cologan, dean of the Beijing diplomatic corps, presented the outline to Prince Qing. Li Hongzhang was indisposed and did not attend. Minister Cologan requested the prince convey the note to the Qing government for an early reply. Upon receiving the note, the Qing government gave its approval immediately and without any alterations.

In the more than nine months from December 27, 1900, when the Qing government approved the "Outline for Peace Negotiations" until September 7, 1901, when the Treaty of 1901 finally was signed, there arose a fierce debate among the Powers over the treaty's details. The chief differences concerned the two questions of punishments for offenders and the indemnity the Qing was to pay.

As regards the issue of punishment, Britain and Germany were generally on one side, demanding the list of offenders be lengthened, and the stiffest possible punishments be meted out. Russia, the United States, and Japan favored reducing the number of offenders and imposing less severe sentences. As early as August 28, 1900, MacDonald telegraphed Salisbury that "the punishment of those taking a prominent part in the recent outbreak against foreigners is of great importance in its bearing on the future. Unless severe punishment is inflicted on individuals . . . it will only be a question of time for a recurrence of the present crisis."[34] His words reflect the position of the British government on this issue. Obviously, the British purpose was to diminish the Russian influence over the Qing government and to take advantage of an opportunity to help pro-British Qing officials come to power. On September 17, in a circular note to the Powers, Germany demanded the Qing government turn over the chief offenders "as a prerequisite for diplomatic negotiations." Under pressure from Britain and Germany the Qing government issued an "edict" on September 25 announcing that Zaiyi, Zaixun, Zailan, Zailian, Zaiying, Pujing, Gangyi, and Zhao Shuqiao were to be deprived of their titles of rank or nobility and to be handed over to the officials concerned for punishment, because they "connived at and shielded the Quan bandits [Boxers] and provoked discord with friendly nations."[35]

However, the "edict" did not satisfy Britain and Germany, for some thought that the Empress Dowager Cixi should be included in the list of chief culprits to be punished. This had to be resolved before other matters. For a time Germany advocated punishing her. After much deliberation, however, Britain believed that it would be advisable to maintain her ruling status, so as to avoid further disturbances or the partition of China, thus endangering British interests in China. In a talk with the German ambassador to Britain, Salisbury stated plainly that "it is impossible to punish the Empress Dowager."[36] Speaking in her defense, Zhang Zhidong, a pro-British bureaucrat, said to Fraser, the British acting consul general at Hankow: "Nor did she during thirty years of power ever show hostility to foreign persons or things; in fact, she welcomes foreign inventions in her own palace, and tried to enter into friendly relations with foreign ladies." Zhang also said emphatically: "As the emperor is Her Majesty's adopted son, and as the Chinese policy rests on the principle of filial piety, His Majesty would not be able to face his people or command their allegiance, if he should allowed his mother to be disgraced."[37] His words were taken seriously and appreciated highly in British official quarters.

Although the imperialist Powers agreed to maintain the Empress Dowager's rule, they differed widely over the punishment of the other offenders. Britain demanded capital punishment for all offenders designated in the joint note of December 24, 1900. This position was supported by France, Germany, Austria, and Italy, but "unanimity was prevented by the refusal of Russia, Japan, and the United States to demand the death penalty for Prince Duan [Zaiyi], Prince Lan [Zailan], and Dong Fuxiang."[38] On February 5, 1901, the diplomatic corps held a meeting to discuss the problem. Pichon said that he had received the instructions of his government in Paris to desist from demanding a death penalty for Zaiyi and Zailan. With France going over to the side of Russia, the United States, and Japan, while Britain, Germany, Austria, and Italy all still insisted on the original proposal, the discussion reached an impasse.

Then, Pichon suggested seeking a compromise on the question of punishments. While supporting the British position, German Minister Mumm believed that the opposition of Japan and the United States meant, "the psychological moment when it might have been obtained passed."[39] Consequently, Satow had to back down. After many consultations the ministers agreed to the following article: "Prince Duan [Zaiyi] shall be condemned to imprisonment pending decapitation. If, immediately after being condemned, the emperor thinks fit to spare his life, he shall be sent to Turkestan, there to be imprisoned for life, no commutation of the punishment being subsequently pronounced against him."[40] The same formula was adopted for Zailan, but for Yingnian, Zhao Shuqiao, and Yuxian decapitation still was demanded. Decisions were reached also on the cases of the other main culprits.

On February 21, 1901, the Qing Court issued a decree accepting all the Allied demands for punishment, with the sole exception that Yingnian and Zhao Shuqiao were "ordered to commit suicide" instead of dying by decapitation.[41] The decisions of the diplomatic corps on February 21 produced the stipulations concerning punishment of offenders included in article II of the Treaty of 1901 and its enclosures IV, V, and VI.

The questions about the indemnity produced the sharpest contradictions and conflicts among the Powers. Generally speaking, Russia and Germany wanted to exact the largest possible indemnity from China because Russia needed money to make up its depleted treasury and to speed the construction of the trans-Siberian railroad, while Germany wanted the money to expand its navy so it could be used in its attempt to reshape the world. On the other hand, Britain, the United States, and Japan proposed the indemnity should be kept within definite limits so that China's potential for imports would not be damaged too greatly. That position served to protect these nations' commercial and other economic interests. Both groups, however, worked as vampires against the Chinese people and differed only in their methods of sucking blood.

The total amount of the indemnity, suitable guarantees, and the methods of payment all produced long disputes among the imperialist Powers. At times their differences threatened to break down the negotiations. Russia proposed that each Power should ascertain its own demands for indemnities and submit them to China separately. Russia objected to appointing a committee to review the demands of the individual Powers. As for the method of payment, Russia advocated that China raise a loan jointly guaranteed by all the Powers. Supported by Germany and France, it also proposed an increase of China's import duties to 10 percent ad valorem to enable China to pay the indemnity.[42] Britain and the United States opposed each Power submitting its own demands and preferred to fix one total sum for the indemnity first and then to negotiate with China. Finally the Powers would settle among themselves on how to divide the indemnity payments. Britain and Japan also opposed any increase in import duties beyond the existing 5 percent ad valorem rate. Their position also included the point that import duties also should not be raised if the *likin*, an internal transportation tax, were abolished.[43]

On March 22 a so-called Indemnity Committee consisting of the British, German, French, and Japanese ministers was appointed to consider the sources from which China could finance the indemnity payments. On March 24 the committee held its first session and decided to invite some "Old China Hands" to provide information on China's revenues. These included Robert Hart, the inspector general of the Chinese Imperial Maritime Customs Service; E. G. Hillier, director of the Hong Kong and Shanghai Banking Corporation; and M. Pokotiloff, director of the Russo-Chinese Bank. Each submitted a separate statement. In addition, the ministers of Britain, Germany, and Japan also submitted their own memoranda. They all agreed that China could afford to pay an indemnity of fifty to sixty million British pounds sterling. Hart's statement was the most detailed, and he proposed that the indemnity be paid in yearly installments, guaranteed by the Maritime Customs Service, the *likin*, the salt gabelle, and the native customs revenues. Import duties were to be increased to an effective 5 percent ad valorem. All the Powers accepted Hart's proposals. Later, the stipulations concerning the indemnity in the Treaty of 1901 were worked mainly on the basis of Hart's statement.[44]

In conclusion, the contradictions and conflicts among the

imperialist Powers in China were extremely sharp and complicated at the time of the Boxer movement. After the signing of
the Treaty of 1901, these differences diminished, but the
scramble for Northeast China between Japan and Russia grew
increasingly fierce in the next few years. It finally led to the
Russo-Japanese War, which resulted in a reshaping of the
spheres of interest in China.

In the past, some people held that there were two different
groups among the Powers at the time of the Boxer movement:
Britain, the United States, and Japan on one side and Russia,
France, and Germany on the other. In fact, that is not quite
correct. Throughout the sixty years following the Opium War,
Britain and Russia were the two major Powers that invaded
China, each seeking to dominate East Asia and hoping to have
China to itself. After the Sino-Japanese War of 1894-95, the
contention for China among the imperialist Powers became ever
more intense. As Russia's ally in Europe, France often supported Russia's aggressive activities in China as a means of up-
holding the alliance. On the other hand, the British attempted
to use Japan to stop the expansion of Russia's influence in
China, while Japan also wanted to rely on Britain's support in
its struggle with Russia for domination in Northeast China. An
alliance of those two countries was in the making.

At the time of the Boxer movement, there was a clear tendency for Russia and France to form the heart of one bloc and
England and Japan to group into a second camp. Germany and
the United States were less clear in their stance. For example,
on the problem of punishing the offenders, Germany supported
Britain, whereas the United States sided with Russia. On the
question of the indemnity, their positions were reversed. Germany supported Russia, but the Americans went with the
British. In short, when discussing the chicaneries and intrigues
of the imperialist Powers against one another and the alternat-
ing collusion and conflict, we need to make a concrete analysis
of concrete problems. The argument that the Powers fall into
two hard-and-fast blocs during the Boxer movement is not in
accord with the facts.

To some extent the contradictions and conflicts among the
imperialist Powers obstructed the realization of their schemes to

partition China and forced them to adopt the policy of maintaining the status quo, so as to avoid an open fight within their own ranks. Some people have stressed that the Boxers frustrated the imperialists' ambition to carve up China. That is certainly correct; nevertheless, we need to include the effects of the contradictions and conflicts among the Powers.

Notes

1. Great Britain, Parliament, Parliamentary Papers (hereafter BPP), China no. 3 (1900), doc. 104.
2. Deguo waijiao wenjian youguan Zhongguo jiaoshe shiliao xianyi (Selected Translations of Materials from the German Foreign Ministry Records on Relations with China; hereafter DGWJ), juan 2, pp. 14-15.
3. BPP, China no. 1 (1901), document 53.
4. Ibid., doc. 54.
5. DGWJ, juan 2, p. 31.
6. Hongdang zazhi youguan Zhongguo jiaoshe shiliao xuanyi (Selected Translations from the Magazine Bolshevik on Relations with China; hereafter HDZZ), p. 225, n. 5.
7. DGWJ, juan 2, p. 48.
8. Ibid., pp. 12, 69.
9. HDZZ, p. 233.
10. BPP, China no. 1 (1901), doc. 215.
11. Sergei Yulyevich Witte, The Memoirs of Court Witte (New York: Howard Fertig, 1967), p. 111.
12. BPP, China no. 5 (1901), doc. 1 enclosure.
13. Ibid., China no. 1 (1901), doc. 205.
14. Ibid., China no. 5, (1901), doc.88, Foreign Relations of the United States, 1900 (Washington, D.C.: Government Printing Office, 1903), p. 278.
15. DGWJ, juan 2, p. 197.
16. Ibid., p. 208.
17. Ibid., p. 209.
18. BPP, China no. 5 (1900), doc. 1 enclosure.
19. Ibid., China no. 1 (1901), doc. 254.
20. DGWJ, juan 2, p. 89.
21. HDZZ, p. 239.
22. BPP, China no. 1 (1901), doc. 322.
23. Ibid., doc. 304.
24. DGWJ, juan 2, p. 291.
25. BPP, China no. 1 (1901), doc. 307.
26. DGWJ, juan 2, pp. 87, 94.
27. Ibid., pp. 91-92.
28. Ibid., p. 90.
29. BPP, China no. 1 (1901), doc. 238.
30. Ibid., doc. 396.
31. Ibid., doc. 396.
32. Ibid., China no. 5 (1901), doc. 66.
33. Ibid., doc. 258.
34. Ibid., China no. 1 (1901), doc. 254.

35. Guojia danganju, Ming Qing danganguan (Ming-Qing Archives of the National Archives), eds., Yihetuan dangan shiliao (Archival Materials on the Boxers; hereafter DASL) (Beijing: Zhonghua shuju, 1959), 1:642.

36. DGWJ, juan 2, p. 126.

37. BPP, China no. 5 (1901), doc. 138.

38. Ibid., China no. 6 (1901), doc. 54.

39. Ibid., doc. 55.

40. Ibid., doc. 233.

41. DASL 2:967.

42. BPP, China no. 6 (1901), docs. 3, 84, 101, 209.

43. Ibid., docs. 13, 34.

44. See Ding Mingnan, "Xinzhou tiaoyue yu diguozhuyi" (The Protocol of 1901 and Imperialism) and "Gengzi Peikuan--yifu diguozhuyi zhengzangtu" (The Boxer Indemnity--a Portrait of Imperialists Scrabbling for Booty), in Yihetuan yundong shi luncong (Collected Essays on the History of the Boxer Movement), (Shanghai: Sanlian shuju, 1956).

LIAO YIZHONG

Special Features of the Boxer Movement*

In recent years there have been significant differences of opinion about certain issues in the history of the Boxer movement. This is good, for it reflects a greater depth of research. I believe that by searching for truth through facts, examining things in depth, and free discussion we can reach conclusions that more truly correspond with reality. In this article I will raise a few points about the nature of the Boxer movement and some related issues.

A just struggle, a primitive organization

In general, the organizational leadership and direction for an uprising or political movement match its political mission and nature. The Boxer movement was a just, anti-imperialist, patriotic struggle that matched the demands of historical development. Still, because of some special historical conditions, its leadership and organization both were primitive. Consequently, it emerged as a movement with a progressive nature, but backward in terms of organization. This backwardness is best seen in the two aspects discussed below.

A fragmented organization without a leadership core

A basic condition for carrying out a struggle is to possess a strong and complete organizational leadership. The Boxers' situation was just the opposite. Because of a long period of cruel suppression by the landlord class, the already fractionalized

*From Yihetuan yundong shi taolun wenji (Collected Articles on the History of the Boxer movement) (Jinan: Qilu shushe, 1982), pp. 161-75. Translated by Jay Sailey.

White Lotus sect became increasingly fragmented after the mid-Qing. This was even more pronounced among the Boxers, who were a branch of the White Lotus.

This condition arose because many secret religious sects and popular secret societies rushed to join the Boxers' ranks when they saw the struggle was one for national sovereignty. This produced a diverse collection of independent organizations without any unity. Even in the case of the most centralized forces, the Tianjin Boxers, there were eight branches derived from the Eight Trigrams (Bagua), plus the Red Lanterns (Hongdengzhao) to make a total of nine elements altogether. Within any one branch there was considerable latitude for independent action. For example, in the Qian Trigrams there were several important factions led by Cao Futian, Wang Chengde, Liu Chengxiang, Pang Wei, Han Yili, and Yang Shouchen. Other small groups and elements were too numerous to count.

Boxers in various locations took villages or towns as their basic organizational unit. A unit would be made up of several dozen or several hundred followers formed into a separate branch. In the movement's early stages, relations between these Boxer units assumed simple forms such as "boxing demonstrations" (*liangquan*), "group visitations" (*baituan*), or "mutual invitation gatherings" (*chuantie xiangyue*). At the movement's height even these forms were not much in evidence. Sometimes the struggles between Boxer groups were self-destructive. In all branches, the organization and membership were self-directed, without any discipline. When the crisis came and it was time to resist the Allied forces, the Boxers at Beijing and Tianjin, contrary to what one might have thought, just melted away or "went back to farming after the rain."[1]

The Boxers were a large-scale, major uprising, but they lacked overall organizational leadership. In the history of agrarian uprisings from examples of Chen Sheng and Wu Guang down to the Taiping Rebellion and even including the White Lotus uprisings, there really is no precedent for this. This means that we could say that the Boxer movement was the most primitive in terms of organization of any agrarian uprising in history.

It is not uncommon that in the initial stages of an agrarian uprising, as with the Boxers, the organization is fragmented. But unlike the Boxers, the other examples of large-scale uprisings usually coalesce rapidly around one branch or several armies. In contrast, the Boxers from start to finish remained disorganized.

Normally, such times produce heroes; the Boxers, though they had some minor leaders and strongmen, had no real heroes. There are some objective reasons for this, such as the fact that the Boxers were a broad, mass movement. Also the Qing government kept close records about them and the Boxers' command was fragmented. But the most basic reason was that the White Lotus was a secret religion. These origins produced the fragmented and faction-plagued Boxers. All this was an immensely complex situation that could not be reversed in a short time. Even though the Boxers had developed rapidly and were extremely numerous, they remained a plate of sand with no force to hold it together.

This fragmented quality produced serious, even calamitious losses. Because the Boxers were fragmented, they could not bring into play their own great strength of numbers, to strike blows against the Qing rule and become a major cause of the Qing collapse. We can say about the Boxers, from the time that the Harmony Boxers (Yihequan) first rose in revolt down to the movement's high point, that the leadership actually fell into the hands of the Qing Court, except in the province of Shandong. As a consequence when the Qing Court changed its policies from pacification to suppression, tens of thousands of Boxers were unprepared to resist the government's surprise attack and were quickly defeated. Even the strongest group, the Tianjin Boxers, were defeated by Song Qing in a single day. The Beijing Boxers, described as "a threat" to the Qing Court's continued existence, were quickly dissolved. When the Allied forces entered Beijing, there were scarcely any Boxers left to resist them. The major reason for this was the Boxers' fragmented nature.

From this we can see that, in spite of its fame and prestige, the Boxer movement was really quite weak. I am afraid that both in logical and realistic terms it makes no sense to consider the Boxers as strong or to deny that they were controlled by the Qing Court.

Some studies divide the Boxers into two types, real and false groups, or sometimes official and private groups. The grounds for drawing these distinctions appear to be the backgrounds of the participants, the degree of their antiforeign sentiments, their participation in violence and looting, or their acceptance of and resistance to pacification. This type of analysis does not reflect reality. The Boxers were an organization based primarily on the peasantry, but their numbers included some landlords and people from other classes within society. It is wrong to distinguish between "true" and "false" Boxer units on the basis of their members' backgrounds. Not all the people who wished to destroy foreigners were in the Boxer movement, while those who were violent and destructive were not necessarily all from outside the Boxers' ranks.

The willingness to accept pacification also is not good grounds for trying to distinguish "true" from "false" Boxer units. In most cases--again with the exception of Shandong where a policy of pacification was not implemented--the Boxers willingly accepted pacification. This includes elements that were quite important such as those of Li Laizhong, Cao Futian, Zhang Decheng, Wang Chende, and others.

When considering this matter of organization, we should note it was particularly after the Boxers developed in the Beijing-Tianjin region that the masses took over, and there was no way of determining who gave approval for things. Anyone could put a banner on a pole and set up a group of Boxers. Consequently, to attempt to distinguish clearly between "true" and "false" groups of Boxers makes no sense, for there was no clear distinction. Others ask if these labels meant the Court was attempting to categorize the Boxers in order to carry out their own bloody suppression, or if the Court's attempts to categorize were intended to promote the Boxers' purity. None of these interpretations makes much sense in light of the facts. Certainly bad elements were mixed in with the Boxers. No uprising in human history has been able to avoid this phenomenon, and the Boxers were no exception. For these reasons we should not draw distinctions between "true" and "false" Boxers.

Deep feudal superstitiousness

The Boxers, in both their thought and their actions, revealed their deeply feudal religious and superstitious nature. The White Lotus sect was a secret religion embracing a mixture of Buddhism, Daoism, and Manicheism. It had existed for centuries among the lower classes of society, and a number of other elements had become mixed in with it. These included wizardry and the occult, complex religious and deeply Confucian concepts. Even the heroes of novels and plays were added. All of this made their teaching eclectic and all-encompassing. Added to this were the oral teachings of the White Lotus sect itself with its magic phrases that could only be spoken, not written down. This approach increased the probability that there would be mistakes in transmitting these teachings or that those who transmitted the doctrine might add religious or other elements of their own. The Boxers' doctrine grew increasingly diffuse until it seemed to include all the beliefs current in feudal society. The Boxers' teachings contained many religious or superstitious elements such as impenetrability to firearms and swords, swallowing talismans and chanting incantations, possession of the body by spirits, and prophesies through spirits or Buddhas. Yet, obviously the battle against imperialism could not be won through a belief in personal imperviousness to bullets and swords. Distinctions between friend and foe could not be based on whether a yellow paper when burned would float in the air. Decisions about great actions could not be controlled by unseen spirits or Buddhas.

I have already noted that the Boxer movement, in spite of its great fame and prestige, lacked unity in practice. Moreover, although it had a truly mass basis, it lacked a leadership core. Although its struggle was just, the ideology and organization were exceptionally backward. It relied on spiritual weapons based only on simple ideas of revenge and could advance its cause only through spontaneous actions.

It should be said that in the initial stages of a popular movement, religious superstition has a certain usefulness in inciting action by the masses and arousing their determination. Nevertheless, religion does not constitute a scientific understanding of the objective situation and thus cannot sustain a long-term spirit of struggle and solidarity among those participating in an uprising. Religious superstition cannot resolve certain problems encountered in the process of any uprising. On the contrary religious ideas push people into the sad plight of losing their direction.

In the economically backward areas of North China of those days there was neither a proletariat nor a bourgeoisie available to lead the peasantry. Because of the way in which China's situation had developed, the anti-imperialist movement's leadership naturally fell to the Boxers, who had a base among the masses. But a secret body such as the Boxers could not produce overnight the organizational leadership and ideology that the situation required. The backward organizational leadership and ideology of the Boxers was not coincidental but was determined by the objective conditions of time and place. The Boxers' feudal ideology and primitive organization certainly could not provide the leadership needed to oppose feudalism. Moreover, it would be given a great blow by the foreign aggressors.

**Two sides of the slogan "aid the Qing
and destroy the foreigners"**

The Boxers' basic principles and basic objective is found in their slogan "aid the Qing and destroy the Foreigners" (*fu Qing mie yang*). It summarizes and represents the spirit of the Boxer movement. This objective, however, has two sides: a positive and a negative one.

This slogan began to be used in the April 1898 Dongchen uprising in Sichuan's Dazu county, and sometimes "aid" became "help," "promote," "preserve," "assist," or "follow" in variant forms. Similar banners were raised in Hubei, Hunan, and Yunnan. In October 1898 the slogan appeared almost simultaneously in Shandong and Zhili; shortly afterward it appeared in Honan. It is worth noting that this slogan was used only in struggles involving Christianity. It did not appear in the struggles against the foreign concessions or in the struggles against Russian and German railway construction.

We also should note that in all the cases from the Dazu uprising through the Boxers, the slogan was not used merely in opposition to foreign religion, but involved a combined opposition to Christianity and the impending partition of China. The manifesto issued by the Dongchen uprising said that it opposed the foreign Powers because they "insulted our dynasty and dominate our officials" and "they want to carve up our national territory."[2] The Boxers pointed out that the "most hateful treaties" had compelled them "to avoid falling under foreign domination or occupation" and required them "to destroy the foreigners." We can see from this that the Dongchen uprising actually was the start of the Boxer movement.

The slogan "aid the Qing; destroy the foreigners" did not imply that the Qing court was considered the puppet of the foreigners and should be overthrown. Rather the slogan takes the Court as something to be supported against the enemy while pursuing "the destruction of the foreigners." This is revealed in the combination of the contradictions between popular religion and nationalism, the nationalist contradiction being the primary one. The contradiction between classes within China fell to a secondary position, because both the upper and lower classes wanted to fight against insults and sought to preserve the nation. This is just as Lenin said: "Every slogan should be raised in accordance with the special circumstances of the specific political situation."[3] The use of the slogan "aid the Qing; destroy the foreigners" was a result of "the special circumstances of the specific political situation." Thus, within a few months, areas north and south of the Yangzi came together without any prearrangement (although sometimes through direct contact) and raised this goal. This was not merely a coincidence; it was the proper path for the Chinese people of that time in their attempts to save the country from disaster.

Even in late Qing times, we do not lack examples of cases like this in which the governed and the governing classes come together to try to save the country. The bourgeois class through the Reform Party sought the support of the Guangxu emperor and Court officials in carrying out its capitalist constitutionalist reforms. Sun Yat-sen, the leader of the democratic revolution, hoped to use Li Hongzhang's power to advance reform and overthrow the Qing dynasty. The Revolutionary Alliance

(Tongmenghui), while it controlled half the country, still sought the support of Yuan Shikai, a representative of the landlord class, in order to overthrow the feudal system and replace it with a bourgeois republic.

The 1898 Reform, the 1911 Revolution, and the Boxer movement were all quite different in terms of philosophical bases and political tendencies. They also differ considerably in practices and results. Nevertheless, they shared a common desire to free China from its semicolonial status and to protect the independence of the Chinese people. To different degrees, each appeased the landlord class. All were alike in seeking to gain the support of elements within the landlord class.

The Reform of 1898 and the Revolution of 1911 both had a shaping influence on our ability to learn from the more developed civilization of the West and in terms of Chinese capitalism's development. Both appeased not only the power of feudalism but also imperialism. Because of this we must consider carefully the Boxers' slogan "aid the Qing; destroy the foreigners," and we have no basis for insisting that their slogan should have been "sweep away the Qing; destroy the foreigners" (*sao Qing, mie yang*).[a]

The underlying problem was that the "Qing" with whom the Boxers were allied were different from the Guangxu emperor, Li Hongzhang, or Yuan Shikai. The "Qing" turned out to be the most backward and obstinate representatives of the landlord class. Perhaps it is this characteristic that prevents some people from seeing the tragedy of the Boxers. Another problem involves how it was that both the bourgeoisie and the peasantry in China, in carrying out their own patriotic programs, were willing in some degree to appease the feudal rulers and seek their support. This is not simply a mistake made by their leaders or themselves, but it is partially a result of the fact that under the oppression of Chinese feudalism and international capitalism, the bourgeoisie was at first too weak and later too imbalanced. The peasants were narrow and limited by nature, with the inevitable result that the bourgeoisie abandoned peasant leaders. To put it another way, the feudal landlord classes of that time were still quite strong.

The slogan "aid the Qing; destroy the foreigners" had a positive value for the Boxer movement in that it helped the Boxers develop. Especially in the early stages, this slogan had the effect of reducing opposition and broadening the movement's anti-imperialist appeal. This was so because among those truly patriotic officials, landlords, and Qing soldiers, the slogan was something they could accept, even though they could not fully implement it. To a certain extent they felt sympathy with and support for the Boxers and with the struggle to "destroy the foreigners." At the same time the various ranks of Qing officialdom were at a loss about what to do. "Sometimes they supported the Boxers, other times they said they wanted to lead them; sometimes they spoke menacingly about complete suppression . . . their policies were never stable."[4] Undoubtedly this situation had a direct influence on Qing government policies and was helpful in the development of the Boxers.

The Boxers' objective of "aid the Qing; destroy the foreigners" also had considerable defects and a negative side. The Boxers had only this slogan, and the rest of their ideas consisted only of a flurry of announcements in which religion and muddled anti-imperialism were thrown together. The slogan was both their highest objective and the real substance of their policy. This slogan was obvious and simple, but it was also vague and ambitious. It was not clear what policy should be employed to "aid the Qing" while also "destroying the foreigners." What Boxer policy would prevent the Qing and foreigners from joining forces to wipe out the Boxers? The Boxers had no answer.

The aim of any struggle must be based on policies and strategies that distinguish and deal with the relations among oneself, one's friends, and the enemy. From the actual record of the Boxers we can see that they neither understood nor could distinguish among the Qing authorities and the "foreign" capitalist things in China on this basis. For example, their support of the "declaration of war" and "pacification" were correct, but in the process they lost their autonomy and became tools of the Qing Court. Another example is that the Boxers supported the Empress Dowager and her lackeys, whose purposes were quite different from their own, but showed great enmity toward someone like Nie Shicheng and his troops, who wanted to fight the invaders. Adding the participation of another faction, the White Lotus Society, with its advocacy of vengeful murder and anti-Qing sentiments, was another major mistake.

Although the Boxers could join with the landlord class to "aid the Qing; destroy the foreigners," they could not cooperate with the bourgeoisie because that class had no desire to assist the Qing and could not accept the principle of "destroying the foreigners." When "destroying the foreigners," the Boxers could not distinguish scientifically among them, but there are many reasons for this. One important factor was the vagueness of the Boxers' objectives. As a consequence of confusing up the differences between friend and foe, the Boxers' ability for struggle was weakened, obstacles appeared that could have been avoided, and, in the biggest mistake of all, they came to be dominated by Court officials from behind the scenes.

Among the more important peasant uprisings, some from the start aimed their struggle directly against the feudal Court, while others came to oppose the Court in the process of a struggle against certain elements of the landed gentry and corrupt officials. Some of these uprisings used slogans of opposition to the feudal land system, but the Boxers did not. All through the course of their uprising, their objective was "aid the Qing; destroy the foreigners." This makes them a special case in the history of Chinese agrarian uprisings. The Qing attempted to pacify the Boxers through a policy of both suppression and support, with an emphasis on the latter. In this manner the Boxers and the Qing Court gradually reduced the contradictions between themselves, which originally had been based on class differences. Step by step the Boxers moved toward the Court; "aid the Qing" became increasingly important.

There is a special reason why the Qing adopted a policy of pacification toward the Boxers. The Court was being manipulated from behind the scenes and actually was quite weak. The Court feared that suppression would stir up a real resistance among the people, so it adopted a policy of pacification as "an expedient."[6] At the same time, it could make use of the Boxers for its own political objectives.

In Shandong the situation was unique. Local officials from

Li Bingheng to Yuxian were all dissatisfied with the spread of Christianity and were anxious about the increasing aggression by the Powers. In addition, these men were all blindly xenophobic. This produced the policies of Zhang Rumei that sought to "enroll the boxing adepts into local militia," or to "make private groups into public ones," or to "turn the boxing braves into local militia."[7] It went to the extent that the Court, "fearing the foreigners themselves, provoked disturbances between the masses and the foreigners."[8] In addition, the Boxers' "aid the Qing" concept did not present a major threat either to the Court or to the local officials. At the same time there were those in the landlord-official class who really advocated resisting the foreigners. The pacification policy permitted the Court to gain the support of these people "by appealing to their sympathies."[9]

Given this situation, plus the intimate relationship between officials such as Yuxian and Li Bingheng and the behind-the-scenes faction at Court whom these officials slavish followed, it is easy to see how the ruling faction at Court and the Shandong officials could make a common cause. Together they followed a policy of tolerating the Boxers, sympathizing and supporting them. Thus Shandong continued the policies begun by Zhang Rumei, a corrupt official dismissed from the governor's office in Shandong, and Yuxian, when he was Shandong governor. Yuan Shikai, who replaced Yuxian, generally suppressed the Boxers, but even he was "bound hand and foot" by the Qing policies.[10] He could only "rely chiefly on setting an example by disbanding them."[11] What he did, "in comparison with the techniques of Yuxian, certainly was not so severe."[12] Actually, Yuan Shikai put to death the most unruly of the Boxers, but that was only after May 1900 when he joined the emerging trend of "mutual protection in the Southeast" among those Qing officials who did not support the Qing Court's Boxer policies. It is wrong, without regard to time or circumstances, to label Yuan Shikai as one who consistently undertook a policy of killing Boxers. In Zhili, Yulu's policy was one of conciliation. Such a policy reduced the obstacles for the Boxers and helped them to develop relatively freely. It was just such a policy that made the Boxers accept the illusions behind the "aid the Qing; destroy the foreigners" slogan, thereby giving the Qing a basis to carry out a policy of pacification after June 1900.

The peasantry, before it has the leadership of an advanced class, cannot possibly throw off the shackles of feudalism. That is why peasant uprisings, no matter how deep their opposition to feudal oppression and exploitation, always wind up simply opposing a bad emperor and supporting a good one. At best there will be a change of dynasty--the overthrow of one feudal ruler for another. In no instances do they oppose the entire system, for opposition to corrupt officials and bad emperors is not the same as opposing feudalism. For the sake of preserving the feudal system, its leaders will oppose wrongdoers too.

The demands of history were that the whole Chinese people band together to oppose the Powers and preserve the nation's independence, but the Boxers often could not resist the reality of feudal exploitation and oppression. The facts show that the Boxers did "aid the Qing," for they did not kill officials and steal their seals of office, instead they were finally turned into a tool that "worked hard for the dynasty." At most there was talk of "attacking Beijing," or of killing "one dragon, two tigers, and three hundred sheep,"[b] or the idea that "aid to the Qing" was only a stratagem.

When we look at the facts, the Boxers never really opposed the Qing at all.[13] A great deal of the anti-Qing, anti-feudal materials really are manifestations of their anti-foreignism. For example, the Boxers opposed traitorous officials, lackeys of the foreigners, those who worked for the foreigners, those who dealt in foreign goods, students who studied abroad, or people who had any dealings with foreigners. The vast majority of these people had no relation at all to the aggressors' running dogs.

I can provide two examples that illustrate this. In the first case, for years the "sharing of grain" (*junliang*) has been given as important historical evidence for the Boxers' hostility toward feudalism. The local gazetteer for Guan county in Shandong says, "they took the name of grain sharers, but actually were opposed to the sect," while the Linqing gazetteer states: "That year there was a famine and the poor people joined together with the boxing bandits and called themselves 'grain sharers.' They tore down telegraph poles and robbed the official posts and commercial boats." In the official records collected in both *Shandong yihetuan anjuan* (Documents on the Boxers in Shandong [1980]) and *Shandong yihetuan diaocha ziliao xuanbian* (Selected Survey Materials on the Shandong Boxers [1980]), we note the sharing of the Christian converts' property, comprising both landlords and nonlandlords. There is no clear record of their distributing the landlords' grain. In accounts of a neighboring area we find a statement by an old man in Zhili's Nangong county who recalled that the Boxers "ate [the stores of] rich households," but there is no way to prove this is reliable. Also we cannot know for sure if the "rich households" refers to the Christian converts or to the households of landlords. If this does in fact mean the landlords' households, then it was unusual. If those conditions prevailed in Shandong, it would have been the same in Zhili. For example, the mass organization at Tankouzhen in Tongzhou prefecture said: "Everyone agreed that we wanted to slaughter all the Catholics in our village and its neighbor; then we would divide their property equally."[14] Thus we see that "sharing grain" and "dividing property" refers to the Christian converts, and the phrase "actually hated the sect" means Christianity and has nothing to do with anti-feudalism at all.

My second example shows that many anti-government posters that have been considered anti-Qing were not really that way at all. A poster upbraiding the Manchu nobleman Yikuang, who was in charge of foreign affairs, said: "You live off a salary from China, but you defy her and aid the foreigners. We must enlighten her, change the heavenly ruler, and return to the proper path for the Great Qing."[15] This is an expression of support for the Qing, not one of hostility toward it. Another poster condemned Zhili Governor General Yulu: "You turn to me, but whom can I turn to? Nine of ten officials are fat. The robbers break in, but are not captured. They run away and who is to blame?"[16] While this is clearly hostility toward corruption and the failure to resist the foreigners, it is not anti-Qing. Of course from the great body of Boxer materials we can find some things that oppose feudal rule, but we should not judge the whole by a few examples nor generalize on the basis of exceptional cases.

Thus we can see that because they were not opposed to feudalism or to Qing rule, the Boxers did not pose a threat to the feudal landlord classes. Because of this the Boxers actually received sympathy and support from some officials. Also because of this the Qing Court changed from a policy of pacification to an even more conciliatory policy. This same characteristic also reveals why the Boxers could be manipulated by the Qing.

A genuinely anti-foreign, patriotic movement

As I have discussed above, the Boxer movement lacked a real anti-feudal nature, but it was genuinely an anti-imperialist and patriotic movement. The content of the Boxers' anti-imperialist struggle was opposition to the Powers' aggression in order "to prevent China from being carved up." In concrete terms they were hostile to the power of the missionaries and resisted the aggressive invasion of the Allied forces in 1900.

Hostility toward Christianity formed part of the Chinese people's opposition to the Powers, and was a special theme in the historical opposition to foreign aggression. Hostility toward Christianity permeated the whole process of the Boxer movement. It was at the heart of the slogan "destroy the foreigners" prior to the invasion by Allied forces.

Since the appearance of classes within human society, religion has been used by the governing classes to serve their own interests, directly or indirectly. In modern times, Christianity had become a special weapon of the international bourgeoisie to render the people numb and to help them carry out their aggression. At the end of the nineteenth century, the capitalist Powers were committing aggression against China by military, political, and economic means. They also used foreign religion as an ideological means to enslave the Chinese people. Even more importantly, Christianity became a means to aid in their aggression. A majority of the missionaries, consciously or unconsciously, served foreign aggression as its fifth column or spies. Consequently, the Powers in their aggression against

China always employed the missionaries as an important force. An American minister to China, Charles Denby, wrote frankly: "The energy of the missionaries and those involved in the missionary endeavor is very important to our national interests." Without their help, the Powers would have been unable to carry out the "'responsibilities" of Great Power diplomacy. A well-known missionary admitted that when the Powers came to China for missionary activities, they "said that it was for religious purposes but did not admit their political reasons."[17] Missionary activity was conducted because it was in accord with the needs of the Powers. As the Powers' aggression increased, the pace of the Christianity's development in China also speeded up. We can take as an example a second-class missionary country like the United States and examine the increase in the numbers of missionaries and missionary "investment" in China and the case will become clear. Table 1 is drawn from C. F. Remer, *Foreign Investment in China* [1933].

Table 1

Date	American Mission Groups in China	American Investment in China Missions	Total Foreign Investment in Missions as % of Overall Investment in China	American Investment in China Missions as % of Amer. Investment in China
1875	15	US$100,000	12.5%	42%
1900	31	US$500,000	20.0%	43%

From this table we can see that the increase in missionaries and missionary investment exceed the overall increases. This illustrates the importance of the missionaries in the process of foreign aggression against China.

Because of the activities of Christianity, by the time of the

Boxer movement, it is apparent it had become "an empire within an empire."[18] At the time some gentry and officials cried out that the power of the foreign religion was growing so rampantly the Empire "would not return to the dynasty's control."[19] The harm done by Christianity came to be the greatest among the afflictions visited on Chinese society. For this reason the Boxers were correct in their opposition to the foreign churches as an important element in foreign aggression toward China.

As the Boxers' struggle against Christianity developed, they threatened the Powers' special privileges in China. The Western bourgeoisie raised a great outcry and assembled armies of aggression to wage an undeclared war to suppress the Boxers. It is quite unusual that eight indifferent nations should band together to launch a war of aggression against one country. Actually it is wrong to say that a plan had been in existence for a long time for organizing such an Allied army from the eight nations. But the Powers were ready to go to war to create new colonies, protect their existing colonies, and preserve the semi-colonial forms of control they had in China. This characteristic was an inevitable result of their development from capitalism to imperialism. This is an example of the maxim "imperialism means war." After the Allied force began its battles, politicians of the various foreign nations, missionaries, and their governments all took advantage of the situation to advance various demands for plundering and partitioning China. Their war of aggression against China was only a continuation of their general policy, since before their own "declaration of war" the Qing Court already had extended protection to the legations, churches, and foreign citizens.

The Boxers, in their war against the Allied army, had many heroes who were unafraid of death. They showed a lofty spirit of sacrifice and patriotism; many Qing soldiers also fought heroically. To protect their nation they did their rightful duty and made great sacrifices. The Boxers and the Qing soldiers fought with primitive weapons in battles both with the Allied forces in the North and the army of Czarist Russia in the Northeast. Many battles were hotly contested. For example, they forced the Western armies under Admiral Seymour to retreat along the initial path of invasion toward Beijing. The battle at the Dagu forts lasted seven long hours, and the defenders killed many of the enemy. After the battle around Tianjin's southern gate there were more than 750 Western corpses; at the front lines in Heilongjiang and Fengtian, the Chinese armies and people staged hard-fought resistance and defeated the Russian armies in several instances. This stubborn resistance by the Chinese armies and people caught the foreign aggressors by surprise. They all cried out "The Chinese are very brave this time. We haven't seen anything like it before."[20] German Kaiser William II despaired that the setbacks of the Allied forces "will make Europeans lose face before Orientals, there is no doubt about it."[21]

After the fall of Beijing, the Allied forces expanded their aggression into Zhili. Then, because of their betrayal by the Qing Court, the Boxers faced suppression by a combination of reactionary domestic and foreign foes. Their circumstances were very difficult, yet they still fought with determination against the enemy in many places.

To summarize, the Chinese army and people in their war against the Allied forces and in the process of preserving the nation showed a fearless spirit that calls up inspiration and sympathy. Their patriotism was not hollow, nor was their bravery an empty phrase, but rather their place in history was written with their own lives and blood!

In modern China, the semicolonial and semifeudal modes of production inhibited growth. This prevented the proper development of capitalism. The transformation of production was the key to promoting China's progressive development. To transform production we had to go through a class struggle and achieve political independence. Only then could we achieve our objective. In the modern era many patriotic people became anxious about the fate of the country and its people. They sought to make their country prosperous and strong through ideas such as "save the country through industry," "save the country through science," or "save the country through education." But their efforts reveal that only through class struggle can the objective be attained. Even though they wanted to devote themselves to the nation's best interests, they really had no hope of success. Consequently, based on the premise that class struggle decides a nation's fate, the Boxers' firm stand against imperialism should receive high marks. We cannot dismiss the Boxers based on evaluation of their usefulness in the struggle to change the mode of production or to establish science in China.

We study history in order to sum up the lessons of experience so that later generations may learn from them. Therefore, while we affirm that the Boxers were anti-imperialist and patriotic, we cannot gloss over their negative side. In the process of their anti-imperialist struggle, the Boxers were blindly xenophobic. In their anti-Christianity they broadened the scope of violence; they also placed the blame for all the bad aspects of their society on Christianity. The old proverb "Heaven doesn't rain and the earth dries up" became explainable in their minds as "because the churches block Heaven." They never drew a distinction between Christianity itself and the Powers' use of Christianity in their aggression against China. Also they could not distinguish between the pious Christians and the aggressive ones in the missionary community. Among the Chinese Christians, they lumped the patriotic people together with the traitorous. They expanded the meaning of "destroy the foreigners" (*mieyang*) to the point that they treated as enemies anyone who read foreign books, had foreign goods, or wore foreign clothing. It went to the extent that they branded students who had studied abroad, such as Qin Lishan, as "scabs" (*ermaozi*). Because scientific technology is related to modern capitalist modes of production, they also treated it as their enemy. Their slogans were "tear up the railroads; cut all the lines; destroy the steamships." They thought by doing this they could "celebrate peace and prosperity throughout the great Qing realm." Obviously this was absurd, the product of their own isolation and xenophobia.

If we are to evaluate the blind xenophobia of the Boxers, we must understand more clearly the reasons why it arose. Only then can we understand the situation accurately. Simply praising or condemning the Boxers' xenophobia without being able to explain the reality of the situation would prevent us from drawing lessons from their experiences. Judging from the facts, this anti-foreignism was no accident; there were historical and practical reasons behind those feelings. First of all Chinese society in those days adhered to a strongly isolationist philosophy. This was a consequence of the ignorance of the ruling class and their policy of keeping the country closed off from the outside. The primitive peasantry, bound for generations to a tiny plot of land, could not break out of their mental shackles. Egged on by Court officials working behind the scenes, the Boxers became increasingly antiforeign. Second, Western science and technology, as it penetrated China, tended to have a positive impact on the development of China's economy. Yet, science and technology did not always serve as a means to advance the economic conditions of the people as a whole, but rather appeared in many instances as a tool by which the Powers carried out colonial plunder. To a certain degree and in some aspects, it did harm the people and destroy the national sovereignty.

Christianity oppressed the people everywhere, thus there were objective reasons for the people's hostility toward it. Still, we cannot take these objective factors as the basis to deny that Boxers were blindly xenophobic. For example, some Chinese writers have said that the tearing up of railway lines and chopping down of telegraph poles was necessary for military reasons in the resistance against the Allied forces. This is incorrect. We should not try to explain away the negative aspects of the Boxers.

There is an important difference between opposing foreign aggression and xenophobia. The former is a just activity to uphold national sovereignty and independence; the latter is a foolish manifestation of isolationism and unbridled vanity. The two have some formal similarities and are related to one another. The Boxers, who were not capable of making scientific distinctions, could nevertheless put the stamp of patriotism on their anti-foreign actions. Consequently, the apparent connections and similarities between anti-imperialism and anti-foreignism led to excesses. "Things will develop into their opposites when they become extreme." This happened to the Boxers.

In modern history, a tendency toward xenophobia often was present in anti-imperialist struggles of the Chinese peasantry, who lacked any scientific knowledge. In the Boxers' case these characteristics appeared suddenly and with unusual intensity. The phenomenon of anti-foreignism among the Boxers should be understood in historical terms. It is a backward and negative factor that could not preserve the nation. That lesson is a profound one that later generations should ponder.

As I have stated, the Boxers were not an anti-imperialist and anti-feudal revolutionary movement, but rather an anti-imperialist patriotic movement. They were firmly anti-imperialist, but blindly anti-foreign. It was a popular liberation movement, but not a conscious one.

Notes

1. Qian Bocan et al., eds., Yihetuan (The Boxers; hereafter YHT). (Shanghai: Shenzhou guoguangshe, 1951), 1:52, 348.
2. Jindaishi ziliao (Materials on Modern History) 4 (1957): 29-30.
3. Lenin's Collected Works (in Chinese), 3:107.
4. Missing in the Chinese original.
5. See my article, "Lun Qing zhengfu he yihetuan de guanxi" (The Connections between the Qing Government and the Boxers), Lishi yanjiu (Historical Studies) 3 (1980): 89-106.
6. Guojia dananju Ming Qing danganguan (Office of Ming-Qing Archives of the National Archives), eds., Yihetuan dangan shiliao (Archival Materials on the Boxers; hereafter DASL) (Beijing: Zhonghua shuju, 1959), 1:46.
7. DASL 1:15.
8. Jindaishi ziliao 2 (1978): 20.
9. DASL 1:53.
10. Tao Zhuyin, Shijiu shiji Meiguo qinhua shiliao xianji (Historical Materials on America's Aggression Against China) (Shanghai: Zhonghua shuju, 1951), 1:233.
11. Jindaishi ziliao 2 (1978): 19.
12. Gengzi yihetuan yundong shimo (Complete Account of the Boxer Movement of 1900), p. 24.
13. See my article Lishi yanjiu 3 (1980): 89-106.
14. Quanshi zalu (Miscellaneous Records for the Boxer Period), juan 14, p. 37.
15. YHT 4:147.
16. Jindaishi ziliao 1 (1957), p. 8.
17. Tyler Dennett, Americans in Eastern Asia (New York: 1922), p. 556.
18. Ibid., p. 485.
19. DASL 1:54.
20. YHT 3:191.
21. Deguo waijiao wenjian youguan Zhongguo jiaoshe shiliao xianji (Selected Materials on Relations with China from German Foreign Ministry Documents), juan 2, p. 12.

Translator's Notes

a. Such a slogan actually was used by the Boxers, but more frequently after 1900.

b. This implies killing the emperor (long), generals (hu), and foreigners (yang).

SUN ZUOMIN

Some Problems in the Appraisal
of the Boxer Movement*

Appraising the Boxer movement has been a controversial issue among historians in China. Under the influence of the so-called revolutionary approach to historical science, the Boxers were glorified as a "conscious, anti-imperialist, anti-feudal, revolutionary peasant movement" for many years. During the ten-year catastrophe [1966-1976], Lin Biao and the Gang of Four deliberately lauded the Boxers to the skies, lavishing praise on them as a movement with modern, proletarian overtones. They went so far as to praise even their negative and backward characteristics, such as feudal superstitiousness and indiscriminate anti-foreignism. These became radical "revolutionary actions" that the Red Guards were expected to emulate. One result of this efforts was to smother research on the Boxer movement.

This wrong appraisal of the Boxers was criticized by Wang Zhizhong in two pieces, "Feudal Obscurantism and the Boxer Movement," in the first issue for 1980 of *Lishi yanjiu* (Historical Studies), and "A Word on How to Appraise the Boxer movement," in the newspaper *Guangming ribao* (Brilliance Daily) on August 19, 1980.** His criticism certainly helped to eliminate the pernicious influence of the ultra-left line of Lin Biao and the Gang of Four and to clarify some of the confusion about the Boxers. Also, his efforts help promote the healthy development of studies about the Boxers. Nevertheless, Wang Zhizhong's articles seem to go to the other extreme in

---

*From Yihetuan yundong shi taolun wenji (Collected Articles on the History of the Boxer movement) (Jinan: Qilu shushe, 1982), pp. 185-205. Translated by Zhang Xinwei with David D. Buck.
**All quotations in this article without citations are taken from one of Wang Zhizhong's two pieces.

evaluating the Boxers, especially when he lays particular emphasis on the barbaric, backward aspects of the movement. His own interpretations give rise to many questions and have attracted the attention of a number of scholars. Here, I would like to present my own views on a few of the problems he has raised in hopes of encouraging further discussion which may result in a correct, or relatively correct, appraisal of the Boxer movement.

Can the Boxers be said to have been "a rebellion in response to imperial decrees"?

Wang Zhizhong writes, in refutation of the ultra-left fallacies of Lin Biao and the Gang of Four: "Some articles published during the Cultural Revolution called the Boxer movement 'a great rebellion that shook the world.' In fact, insofar as its high-tide stage is concerned, it was 'a rebellion in response to imperial decrees,' a clear instance in which the Chinese peasantry was hoodwinked by the feudal rulers. The Boxers were a traditional peasant movement that had not cast off the yoke of royalism, even in their struggle against imperialism. The peasantry cannot represent themselves in such a struggle. Rather, such feudal diehards as Cixi, the Empress Dowager, became 'their master, an authority over them.' " In my opinion, Wang's formulation is not appropriate.

In the first place, the meaning of "a rebellion (*zaofan*) in response to imperial decrees" is ambiguous. It is clear enough that the imperial decrees involved refer to those issued by Cixi, but who was the target of the rebellion? It obviously could not have been the Qing dynasty, let alone Cixi. She never would have issued a decree ordering the Boxers to start a rebellion against herself or the Qing regime under her control. From what Wang Zhizhong says later it can be inferred that the Boxers' target was imperialism, but that is also an incorrect conclusion, as I will show later. A more correct formulation of Wang Zhizhang's position could be made by calling the Boxers "an anti-foreign campaign in response to imperial decrees."

Wang Zhizhong's formulation that the Boxers were "a rebellion in response to imperial decrees" is full of self-contradictions. In his newspaper essay, he argues that his formulation holds only "insofar as its high-tide stage is concerned." He explains: "Although I am a young beginner, I am not so foolish as to call the whole Boxer movement 'a rebellion in response to imperial decrees,' and thereby disregard the fact that the Boxers fought against the Qing troops in the movement's initial stage and were bloodily suppressed by the Qing in the last stage." Here what Wang Zhizhong means is that movement was "a rebellion in response to imperial decrees" only during its high tide, but not in its first and last stages. Yet, earlier in the same essay he says, "They [the Boxers] could not represent their own interests in a struggle against imperialism." Is that not self-contradictory? Moreover, he quotes from Marx's *The Eighteenth Brumaire of Louis Bonaparte* to provide authority for his point about how a traditional peasant movement such as the Boxers cannot "cast off the yoke of royalism, even in the struggle against imperialism." The theoretical point Wang Zhizhong is making is correct, but he has misquoted Marx's words in making his point. Marx's original words were: "They cannot represent themselves, they must be represented. Their representative must at the same time appear as their master, as an authority over them." We all know that Marx was saying that the French peasantry was "incapable of enforcing their class interests in their own name," and that "they must be represented" and "protected against the other classes and sent rain and sunshine from above." Therefore, "the political influence of the small-holding peasants finds its final expression in the executive power subordinating society to itself." Thus the French peasantry's royalism helped bring about the restoration of the Bonaparte dynasty.[1]

Marx does not mean that the peasantry "cannot represent themselves in their struggle against imperialism," for the peasants' opposition to imperialism is not related to their political influence over the executive power in society. The truth is, as Wang Zhizhong stated earlier, "Cixi and her ilk would never have tried to wipe out the foreigners in earnest." Their antiforeignism "was only a temporary expedient, while their consistent policy was to safeguard the throne even by betraying their country." How can such a capitulationist clique be called a "representative" of the Boxers in the struggle against imperialism?

Apparently Wang Zhizhong also is aware of this contradiction. To account for it he gives a special explanation in his newspaper essay: "What is 'a rebellion in response to imperial decrees'? Actually this refers to how the Boxers' activities during the high tide of the movement were manipulated, tacitly approved, and supported by the feudal diehards. That is an undeniable fact. . . . I am only using a new expression to refute the hyperbole of the recent past." This explanation, however, is still unconvincing. The verbs Wang uses--"were manipulated," "tacitly approved, and supported"--are all passive, carrying a sense of being fooled or forced, whereas the expression "in response to imperial decrees" is active and conscious, suggesting willingness to serve as obedient subjects and lackeys. This is no mere change in wording; it is a change in the basic characterization.

Practice is the sole criterion of testing truth. To uncover the truth we had better look at the facts. First, only some of the Boxers in the Beijing-Tianjin region were hoodwinked by the Qing diehards during the movement's high tide. Those in other areas persisted in their struggle against the Qing rulers. For example, the so-called old units (*laotuan*), headed by Zhao Sanduo and Yan Shuqin, had a force of more than 20,000. They operated openly in southern Zhili, "now massed, now dispersed; now here, now there." "They could be found everywhere and were hard to track."[2] A band of Boxers headed by Sun Yunrong kept on fighting uncompromisingly against the Qing and killed an expectant magistrate and the battalion commander, Cha Rongsui, in the battle at the Jade Emperor Temple in Jiyang County,[3] thus delivering a heavy blow against the Qing troops. Second, the Qing Court was unable to control all the Boxers in the Beijing and Tianjin area. At Fengtai, the Boxers determinedly stopped Natong and Xu Jingcheng, who were trying to abscond from Beijing under Cixi's orders to collude with Admiral Seymour, commander of the Allied forces. The Boxers solemnly declared, "We Boxers obey only the orders of our elders; not those of the Court."[4] In Beijing the Boxers caught a spy, Na Jicheng, an official of the second rank.

"Prince Duan [Zaiyi] wanted to hand him over to the Board of Punishment, but the Boxers said, 'Your highness obeys the Emperor, but we obey the Jade Emperor.' . . . The Prince had no alternative but let them kill the prisoner." There was a big altar (*tan*) at the mansion of Prince Zhuang, Zaixun, who had been appointed to command the Boxers: "Although the Boxers would send their captives to the mansion as ordered, the Prince lacked the authority to render judgments about their cases. He had to ask the elder (*dashixiong,*) who would burn incense and ask for guidance. A captive could be executed only after the Elder had pronounced the death sentence. Prince Zhuang had no power to intervene."[6] Clearly, it is both illogical and at variance with the facts to claim that the Boxers were merely "a rebellion in response to imperial decrees."

In his attempt to deal with his own contradictory position on this issue, Wang Zhizhong wrote in his newspaper essay: "Did the Boxers lack revolutionary character just because they responded to imperial decrees and were manipulated? Their revolutionary character really lay in their anti-imperialist struggle. At its high tide the movement had two distinct aspects: on the one hand, it was being manipulated when it 'responded to imperial decrees'; on the other, it was resolutely anti-imperialist. Are these two aspects completely irreconcilable? The facts in this case are complicated. The Boxers were only peasants living at the turn of the twentieth century. As such, they possessed a revolutionary character from their anti-imperialism, but were not awakened enough to refuse to respond to imperial decrees. They were not as enlightened as are the Communist Party members of today. In my opinion, it is completely inappropriate to 'modernize' these peasants from the past." Wang's arguments here are also unacceptable. While vainly trying to justify his own assumptions, he reveals more contradictions and problems. In his newspaper essay he asserts "the revolutionary character of the Boxers lay precisely in their anti-imperialist struggle." This implies they did not oppose feudalism. That issue remains under discussion among historians, and there are a variety of views. But just before saying the Boxers are revolutionary for their anti-imperialism, Wang says the same Boxers fought against the Qing dynasty in the initial and final stages of their movement. Don't those efforts count as having "a revolutionary character," and weren't they against Qing feudalism? Certainly we need to avoid the hyperbole of the recent past, but it would be just as wrong to go to the other extreme by keeping silent or negating qualities of the Boxers that should be praised, such as instances of their opposition to feudalism.

To take another example, Wang Zhizhong believes the Boxers could not "represent themselves in their struggle against imperialism." This statement is hard to accept. We all know the Boxer movement was a result of the aggression and oppression by the imperialists and their lackeys. The Boxers intended "to organize themselves into militia in order to protect their persons and property." Even foreign commentators of the time understood this: "Recent developments in North China actually resulted from the repeated, unjustified interference of the European countries, which aimed at nibbling away China's territory. China had been suffering from it for a long time." Therefore, the Boxers, "took the opportunity to stir up trouble so as to throw off the oppression of the foreigners."[8] Because of these characteristics the Boxers of other places responded promptly when Zhao Sanduo and Yan Shuqin raised the standard of revolt; rebellion spread like a prairie fire. The historical facts show the Boxers rose for their own interests and those of their nation. Therefore Wang Zhizhong's assertion that "they could not represent themselves in their struggle against imperialism" is wrong. Instead of admitting his mistake, Wang Zhizhong goes on to argue that "feudal diehards like Cixi" became the Boxers' representative in fighting imperialism! We cannot help but ask, who did the Boxers represent if they did not represent themselves when responding to the Court decrees at the high tide of the movement to attack the foreigners? The question seems difficult for Wang to answer.

Wang is wrong also to argue that the Boxers lacked sufficient awareness to refuse to obey imperial decrees. Wang knows that Boxers such as Zhao Sanduo, Yan Shuqin, and Sun Yunrong, whom we mentioned before, were peasants who refused to respond to imperial directives and kept up their resistance to the Qing dynasty. Thus it appears that Wang has no grounds for claiming the Boxers "were not awakened enough to refuse to respond to imperial decrees." This in turn would make Wang's conclusion that some historians have arbitrarily "modernized" the peasants in the Boxer movement incorrect and even seeks to "obscurantize" them completely.

We can see from this discussion that the study of history is a science. Formulation of any historical thesis must be based on the fundamental tenets of Marxism supported by ample, reliable historical evidence and should not be made just for the purpose of stating something new and different. If otherwise, historians will not contribute to the study of the past and will create inextricable contradictions. This should be a lesson to all of us.

On the issue of "an inert force in history"

Most historians agree that the Boxers exhibited a strong, indiscriminate antiforeignism, but they differ on how to evaluate this quality. Wang Zhizhong takes both sides. First, he affirms the Boxers' antiforeignism: "The outbreak of the Boxer movement was a result of the contradictions between imperialism and the Chinese nation. It was not what the imperialists and their lackeys call an unprovoked insurrection by a mob. Theirs was a just revolt that was rational and historically inevitable." But then he hastens to add: "No doubt, their indiscriminate antiforeignism was closely linked with their struggle against imperialism. But to carry on their struggle to the extent of expunging everything foreign was an expression of backwardness and not in agreement with the fundamental interests of the Chinese people. Such indiscriminant anti-foreignism becomes 'an inert force in history.' My grounds for making these assertions are that because of feudal obscurantism, the Boxers were extremely hostile to the introduction of new modes of production into China in addition to their opposition to imperialist aggression. Therefore, in the process of resisting the foreigners, they knocked down all new modes of production and all new ways of life that were alien to feudalism. . . . Railways, trains, electric lines, and machines were among their first targets of attack. These attacks were not always based on military requirements." This is yet another point on which I cannot agree with Wang Zhizhong.

As small producers conditioned by the dispersed, backward, and self-sufficient mode of production prevailing in feudal society, the peasantry, generally speaking, tends to have conservative, backward ideas and instinctively to reject new modes of production. Consequently, there is some justice in saying they were an "inert force in history." Still, the basic requirement of Marxist materialist dialects is "to make concrete analysis of concrete conditions." According to this principle, we should pay attention not only to the common characteristics of things, but also to their specific characteristics. We should see not only their general qualities, but also their respective particularities. Otherwise, if we should substitute general for particular characteristics we inevitably will make mistakes.

The basic support for the Boxers came from the peasantry. Undeniably, therefore, they possessed a conservative, backward mentality that rejected new modes of production. We must point out that these peasants lived under the particular conditions of semifeudal, semicolonial society, rather than a typical feudal society. Especially following the Sino-Japanese War of 1894-95, the foreign Powers came one after another to invade China and wreak havoc. "They bullied and humiliated our country, occupied our territory, trampled on our people, and extorted our wealth."[9] China faced imminent partition. In such circumstances, the indiscriminate antiforeignism of the Boxers becomes a just revolt set off by the increasing contradictions between imperialism and the Chinese nation and not as an ordinary case of Ludditism.

The Boxers were unable to make a distinction between the aggressive nature of imperialism and the advanced modes of production, including the science and technology, brought to China through the means of imperialist aggression. For example, the railroad certainly is a product of advanced science and technology, but under specific circumstances railroads assumed the functions of a "means of aggression." The imperialists of that era admitted this openly: "Wherever a new railway goes, power will follow. All rights, including defense, commerce, mining, and communications, will be controlled by the rails and nobody will dare to do anything to us. The rails are as vital as blood vessels in a human body. With them, we will have all the powers. With all the powers, we could order the local officials about and their people will be like meat on our chopping block."[10] Because imperialism used advanced science and technology, including the railroad, as a means of aggression, what they brought to the Chinese peasantry was neither progress nor improvement, but destruction of lives and property. The peasants, who had suffered so much rejection, destroyed these modern devices without distinction. Still, the peasantry's behavior should not be censured as rejection of advanced modes of production because of the peasantry's backwardness and ignorance.

Wang Zhizhong tries to prove with a long quotation that "the peasants were dead set against the railway, not because it was used by the imperialists as a means of aggression, but because the peasants were afraid that the crops alongside the railroad would die." He believes that the peasants only worried about their crops and had no idea the Germans had grabbed vital rights and interests in mining and transportation.

Wang also argues that the Boxers' destruction of railroads, trains, telegraph lines, and machinery "was not always necessary from a military standpoint." In my opinion, this conclusion is not based on historical facts. To begin with, there are few recorded examples of the Boxers conducting Luddite actions. On the contrary, industrial enterprises equipped with modern machinery, such as a flour mill and a match factory in Tianjin, remained intact. Second, although the peasants along the railway "found later that the crops did not die," they still "opposed it to the death" instead of "admiring the foreigners." Third, the peasants' destruction of the railroads, trains, and telephone lines was truly needed for military reasons.

Let's take some examples: "Fearing the Russians would arrive soon, the Boxers destroyed the Tianjin railroad in order to stop them."[11] After the Boxers destroyed the railway stations at Liulihe and Zhangxindian they were asked, "Why did you destroy the railroad and burn the stations?" The answer given was, "To stop the enemy."[12] "After the Allied troops had left for Beijing, wires along the route were often cut. . . . Communications were interrupted."[13] These three examples bear out my point that the Boxers often acted for good military reasons. Of course, I cannot guarantee that there are not exceptions buried in the relevant records. There might be a few cases where the destruction was not a question of military necessity.

Wang Zhizhong demands too much of the Boxers. If their activities were not "always needed for military reasons," then he considers them an expression of indiscriminate antiforeignism or an "inert force in history." By these standards even the forces led by the Chinese Communists could be classified as "an inert force in history" for they could not be absolutely certain that their destruction of railways and telephone lines was always necessary for strictly military reasons. The proletariat in the early stages, when they destroyed machines and workshops, also could not be certain what the effect of their actions would be. Yet, I doubt the author would agree either to call those actions "indiscriminant antiforeignism" or to refer to them as acts of "an inert force of history."

Wang Zhizhong also advances arguments about the effects of the active forces in history by quoting and elaborating on some statements by Marx in "The British Rule in India" and other articles:

While invading the Oriental countries with gunboats, imperialism inevitably brought them new modes of production and new ways of life. Marx calls this role of imperialism "the unconscious tool of history." On the one hand it carried out a mission of destruction by destroying the old Asiatic society. On the other hand, it accomplished a mission of construction by laying the material foundation for a Western style society in Asia. This material foundation is a prerequisite of fundamental importance for the backward nations of Asia if they are to eventually free themselves from aggression, achieve independence, and join the ranks of the advanced countries. . . . Such a material foundation is a prerequisite of fundamental importance for modern China in its efforts to free itself from stagnancy and enslavement. Yet, changes in that direction met with the Boxers' fierce opposition.

I believe the cause of this phenomenon is that new modes

of production came to China along with the gunboats of the aggressors. Nevertheless, in order to free herself from the aggression of these foreign gunboats, it was essential to let those imported new modes of production take root in China. Only then could a new class and a new revolution emerge; only then could China accomplish her historical mission of independence.

It is necessary to quote Wang Zhizhong at such length here because his arguments involve an important theoretical problem: How we are to evaluate the contribution of imperialism in "bringing civilization" to oppressed countries. Although Wang's arguments are incorrect, still they have won approval from some people because of their theoretical coloration. That makes it all the more necessary to analyze his arguments as carefully as possible.

No doubt, the laws and theses summed up by Marx in light of British rule in India are completely correct. They hold good for all Asian countries, including China. While this is true in a general sense, these laws and theses cannot be applied mechanically to particular cases, but must be combined carefully with that country's particular characteristics. Otherwise, we will have substituted general principles for particularity, thus drawing incorrect conclusions and committing serious mistakes.

Wang says that the introduction of new modes of production was "an essential condition for modern China to free herself from stagnancy and enslavement." Generally speaking, this is correct, but it is inappropriate to stretch that general point so far as to assert, as he does, that "it was essential to let the new modes of production take root in China. Only then could a new class and a new revolution emerge." What Wang calls "new modes of production" are nothing but capitalist modes of production. As everyone knows, however, "as China's feudal society had developed a commodity economy, and so carried within itself the seeds of capitalism, China would of herself have developed slowly into a capitalist society even without the impact of foreign imperialism."[15] This fits what actually happened in the past, for during the century from 1520 to 1620, the seeds of capitalism germinated in the Jiangnan region where there appeared the class of townspeople, precursors of the bourgeoisie. If this development had been allowed to follow its own course, it might have produced "a new class and new revolution" in which "China might have accomplished her historical mission of independence." Then there would have been no question of "letting imported new modes of production take root in China." The destruction caused by the advent of the Qing dynasty and the subsequent aggression of imperialism delayed the course of historical development. The first and foremost requirement for China in the modern period therefore was not "to let imported new modes of production take root," but rather to drive imperialism out of China and clear away the impediments that had held back the new modes of production.

Didn't Marx say, some people may ask, "England has to fulfill a double mission in India: one destructive, the other regenerating--the annihilation of old Asiatic society, and the laying of the material foundations of Western society in Asia"?[16] Also, didn't Comrade Mao Zedong say: "Penetration by foreign capitalism . . . apart from it disintegrating effects on the foundations of China's feudal economy, this state of affairs gave rise to certain objective conditions and possibilities for the development of capitalist production in China" thus "hastening the growth" of the commodity economy in town and countryside?[17] Yes, what Marx and Mao Zedong said is indisputably true, but we must have a sound understanding of why imperialism invaded the backward countries. Marx clearly points out: "The ruling classes of Great Britain have had, till now, but an accidental, transitory, and exceptional interest in the progress of India. The aristocracy wanted to conquer it, the moneyocracy to plunder it, and the millocracy to undersell it."[18]

Even in his analysis of tsarist Russia's aims in building railroads in China, Count von Waldersee, commander of the anti-Boxer forces of the Allies, pointed out sharply: "According to my observations, Russia's intention is to leave China in a permanent state of weakness and under permanent Russian control. It would be most absurd to say that she intends to help strengthen China by developing the latter's economy."[19] The Count was talking specifically about Russian imperialism, but all imperialist countries, including, Germany, are jackals from the same liar. Just as Comrade Mao Zedong summarizes: "It is certainly not the purpose of the imperialist powers invading China to transform China into a capitalist China. On the contrary, their purpose is to transform China into their own semi-colony or colonies."[20]

Some people may say that a bad subjective motive may have a good objective effect, although motive and effect are, as a rule, in agreement. For example: "England, it is true, in causing a social revolution in Hindostan, was actuated only by the vilest interests and was stupid in her manner of enforcing them." But "she was the unconscious tool of history in bringing about that revolution."[21] Isn't Marx telling us here that imperialism may produce good effects? Yes, it must be admitted that deviation of objective effects from subjective motives does exist. A bad motive may sometimes produce a good effect, because the dialectic of history is independent of man's will. Therefore, "when you have once introduced machinery into the locomotion of a country that possesses iron and coal, you are unable to withhold it from its fabrication." Furthermore, "you cannot maintain a net of railways over an immense country without introducing all those industrial processes necessary to meet the immediate and current wants of machinery to the branches of industry not immediately connected with the railways."[22]

The introduction of such a huge productive force as the material basis of a new society necessarily will create the material conditions for the emancipation of the oppressed and enslaved peoples. Still these "material conditions" are not the same thing as "the emancipation of the people." "The existence of such material conditions in isolation will neither emancipate nor materially mend the social condition of the mass of the people, depending not only on the development of the productive powers, but on their appropriation by the people." Take the India of that time as an example: "The Indians will not reap the fruits of the new elements of society scattered among them by the British bourgeoisie, till in Great Britain itself the new ruling classes shall have been supplanted by the industrial proletariat, or till the Hindos themselves shall have grown strong enough to throw off the English yoke altogether." So, on the other hand, Marx points out, "At long last, this constructive work has already started" and "we safely expect to see . . . the regeneration of that great and interesting country." Yet at the same time, he points out that this will happen only in "the remote future."[24]

The presence of imperialist control in India and its absence in Japan is the basic reason why Japan joined the ranks of the developed countries soon after the Meiji Restoration [1868], while India remained a backward country for a long time, in spite of the fact that the material conditions for creating a new society had been brought to India with the invasion of British imperialism. Consequently, although England had a double mission in India, it remained a destructive force in the final analysis. As Marx comments: "The historical pages of their rule in India report hardly anything beyond that destruction. The work of regeneration hardly transpires through a heap of ruins."[25]

Based on the above analysis of the actual historical conditions, I come to a very different conclusion than Wang Zhizhong. Although the Boxers' indiscriminate antiforeignism was an expression of their ignorance and backwardness, it also was imbued with an intensely anti-imperialist patriotism. At a time when the imperialist powers were plotting to carve up China, when the danger of national subjugation or genocide was imminent, and when averting this disaster became the most urgent task, it would be unreasonable to demand that the Boxers distinguish between the aggressive nature of imperialism and the advanced modes of production the imperialists brought along as a means of aggression.

At the same time we should not magnify the Boxers xenophobia as "a rejection of everything foreign" and label them an "inert force of history not in agreement with the fundamental interests of the Chinese people" as Wang Zhizhong does. We should remember that in the past century it was the imperialists who "determined the road of the Chinese people to independence and freedom would be long and painful" and not the Boxers' antiforeignism. Imperialism, in collaboration with the Qing feudal diehards, suppressed the resistance of the people, including the Boxers, and turned China into a semifeudal, semicolonial country of a nature that hindered the normal development of capitalism there.

From this perspective, the Boxers' xenophobia was admittedly an expression of ignorance and backwardness but also was objectively "in agreement with the fundamental interests of the Chinese people." If, however, we take the approach used by Wang Zhizhong and say that the peasants who joined the Boxers in resisting imperialist aggression and enslavement "could not become an active motive force in history" or if we conclude that their xenophobia constituted "resistance to new modes of production and new ways of life," one would have to accept the logical inference that imperialism is the disseminator of "civilization" and can bring "new modes of production" to backward countries who should let these "imported new modes of production take root," so that their "historical mission of independence" might be achieved. In a word, they should welcome and be grateful to their colonizers rather than hate or resist them.

What an absurdity! How to evaluate this "civilization" that imperialism brings to the oppressed countries is a serious theoretical question with great, practical significance. But in this field, the imperialists habitually confound right and wrong. Let's take Russia as an example. Russia wantonly vilified the Boxer movement, "raising howls about the savage yellow race and their hostility toward civilization and Russia's task of enlightenment."[26] Russia hailed the bloody suppression of the Boxers as "a victory of European culture over Chinese barbarism" and a "fresh success of Russia's mission as 'disseminator of civilization' in the Far East."[27]

During China's War for Liberation [1946-1949], the American diplomat Dean Acheson sang a similar imperialist tune: "These outsiders brought with them [to China] the unparalleled development of Western technology, and a high order of culture which had not accompanied previous foreign incursions into China." He made this the "second reason why the Chinese revolution occurred."[28] Any muddleheadedness on our part about this question certainly will help the imperialists take advantage of us. Their theories about the contributions of imperialism are harmful to us.

How "the difficulty in explaining primary and secondary aspects of a mass movement" applies to the Boxer movement

A controversy exists over the fact that foreigners and Chinese Christians were killed during the Boxer movement. Wang Zhizhong's view is, "It would be an oversimplification to say that all who were killed were aggressors or Chinese scoundrels," but he does affirm "the justice in the punishment Boxers meted out to those missionaries who were the forerunners of aggression as well as those Chinese Christians who helped do evil." Then Wang quotes from some records that prove that "a considerable number of those killed were innocent." In commenting on this, Wang writes: "Marx says that the 'undignified, stagnant, and vegetative life' of the Asian peasantry 'evoked on the other part, in contradistinction, wild, aimless, unbounded forces of destruction.' These words seem applicable to the Boxers, too." Thus, Wang concludes: "The xenophobia and obscurantism of the Boxers are difficult to explain based on the theory that mass movements have primary and secondary aspects." I find this conclusion even harder to accept.

Did the Boxers kill innocent people when they were punishing the foreigners and those native Christians who had done evil things? There is much exaggeration and even fabrication in the historical record about the Boxers, most of it written by feudal scholars and imperialists. I agree with Wang that the Boxers' excesses were hard to avoid; our differences are that Wang concludes the Boxers were "a wild, aimless, unbounded force of destruction," while I think that their excesses should not obscure their revolutionary character and they should not be considered a "force of destruction" on account of them.

To explain this point I must make clear under what circumstances the Boxers "killed innocent people." As everyone knows, the Boxer movement was triggered by aggression and oppression from the imperialists and their accomplices--missionaries and native Christians. "Those lordly foreign exploiters" recruited natives for the Church and used them as henchmen and lackeys to bully and oppress the local people. They even committed their own atrocities. According to the records, "dozens of jars of human eyes, hearts, livers, and genitals were found in the churches. There were even people who were skinned alive and pregnant women who had their bellies slit open." Especially after the invasion by the Allied forces, the foreigners looted, burned, and massacred. In Tianjin, they massacred common people "with volleys of rifle fire." "Corpses piled up several feet high and covered a distance of miles from the Drum Tower in the city to the Water Pavilion outside the city's northern gate."[30] After the foreigners captured Tongzhou, they "pillaged and raped from house to house. Those who tried to resist would be shot or bayoneted at once. Sixty percent of the city's population were killed."[31] Entering Beijing, they met a group of Chinese, among whom were Boxers, soldiers, and civilians. The invading troops "machinegunned them for ten or fifteen minutes until none of them remained alive."[32] Such accounts can be found readily in the records, and the above sample will suffice to prove that the Chinese people became enraged by the cruel oppression and inhuman massacres by the imperialist aggressors and their lackeys, and so they "rose with a tremendous force like a river breaking through its dykes."[33]

Even a feudal scholar of those days said: "The more the people were wronged, the stronger their resistance became. So, when they had the opportunity to revenge themselves upon the native Christians, they naturally took it out on the churches too. They would not be satisfied until all were destroyed. That is the usual meaning of vengeance."[34] Even the minister of foreign affairs of an imperialist power had to admit that "one can infer from the outbreak of this fierce antiforeign movement . . . of the Chinese that the foreigners made mistakes in treating them at ordinary times."[35] Thus, it is understandable that the Boxers, stirred by intense and deep-seated hatreds, would commit some excesses when punishing their enemies because these enemies previously had slaughtered the Chinese people.

Still, it does not follow that we approve of "killing the innocent," nor do we "vest whatever the Boxers did with the characteristics of patriotism," nor do we "encourage indiscriminate retaliation." The Boxers' "killing of the innocent" was wrong after all, even though it was a product of the "accumulated rancor, which made the revolt soon get out of hand, once it broke out."[36] We should not gloss over, let alone praise, such xenophobic retaliation.

What I object to is Wang's conclusion that the Boxers were "a force of destruction" just because they had killed some innocents. According to the tenets of dialectical materialism, the nature of a thing is determined by the principal aspects of its innate contradictions. Since Wang admits that "the basic cause of the outbreak of the Boxer movement was undoubtedly the aggression and plunder of imperialism, and the immediate cause of most conflicts was the barbarous oppression of the missionaries and the bullying of the native Christians," these should determine the revolutionary and just nature of the Boxers struggle. Compared with this principal aspect, their killing of innocents and their feudalistic ignorance become minor, secondary aspects that did not affect the revolutionary character of their movement. These secondary characteristics cannot change the nature of the Boxers into "a force of destruction."

In addition, according to Marxism's basic tenets, the evaluation of a mass movement should be based upon its nature rather than on the character of the class or the people who started and led it, or even on the way the mass movement was carried out. For example, the Afghan war of independence was led by the king; yet, Stalin affirms its revolutionary character when he says: "The struggle waged by the king of Afghanistan for her independence objectively was a revolutionary one because it weakened, disintegrated, and destroyed imperialism, in spite of the monarchical views held by the king and his comrades-in-arms."[37] Another example comes from the way in which the imperialists characterized the Chinese people's resistance to aggression as "cowardly, barbarous, and atrocious." Engels objected to their "moralizing on the horrible atrocities of the Chinese." He commented: "We had better recognize that this is a war *pro aris et focis*, a popular war for the maintenance of the Chinese nationality. . . . And in a popular war, the means used by an insurgent nation cannot be measured by the commonly recognized rules of regular warfare, nor by any other abstract standards, but should be judged by the degree of civilization only attained by that insurgent nation."[38]

From my analysis, it is clear that Wang is wrong to conclude that the xenophobia and obscurantism of the Boxers are difficult to explain by means of the theory that distinguishes between the principal and secondary aspects of a mass movement. My logic is obvious: The Boxers' xenophobia and obscurantism are but the means of resistance after all, while the nature of the Boxers was "a just revolt" that "had its historical inevitability and rationality. It was not a fazzia as the imperialists and their lackeys call it."

If Wang Zhizhong agreed with the basic tenet of Marxism that the decisive factor is the nature of a thing rather than its means, he could not say that the Boxers were "difficult to explain on the basis of a theory of mass movements that has principal and secondary aspects." For Wang's outlook not only goes against the basic tenets of dialectical materialism and contradicts Wang's own assertion that the Boxers started "a just revolt," but he also fundamentally negates the Boxer movement. He should have considered this more carefully before coming to such a conclusion.

On "the courage to speak the truth" and other matters

Why are there so many obvious problems and contradictions in the formulations of Wang Zhizhong? First, while correctly advocating "seeking truth from facts in historical research" and "the courage to speak the truth," Wang Zhizhong commits the mistake of emphasizing courage while not fully speaking the truth. In his opinion what appears in historical records, even if it was written by feudal scholars or even by imperialists, is "the truth" and therefore should be "spoken without reservation." It is from such material that Wang quotes extensively to provide evidence of the negative and backward aspects of the Boxers.

To prove the Boxers' xenophobia, Wang uses quotations such as: "They hated foreign goods most of all . . . and would destroy both goods and the people who were found selling them." Or "They even killed an entire family of eight just because of a single match." As evidence of the Boxers' zeal, they changed "the names of things to show a clean break with everything foreign (*yang*). 'Foreign medicine' became 'native medicine'; 'foreign goods' became 'native goods'; 'Japanese rickshaw' was altered into 'safety rickshaw'; 'foreign money' became 'the devil's tender'; 'foreign guns' became 'the devil's firearms'; 'railways' became 'centipedes'; 'electric wires' became 'thousand-*li* poles.' " Such comments do appear in historical records, but a large number are fabrications and vilifications that should not be taken as valid. The exaggeration in these records is so obvious that it can be detected by anyone with a discerning eye. In addition, some of these remarks are quoted out of context, so their meaning is altered. For example, before the quotation "they [the Boxers] even killed an entire family of eight," the original text reads: "[the Boxers] called those who believed in Christianity or were involved in foreign affairs secondary barbarians, tertiary barbarians, or even quaternary barbarians, in accordance with their jobs." This indicates that the eight people were killed not because of a single match but because they were lackeys who helped carry out the evils of imperialism.

Obviously, it is wrong to take all records from the past as "the truth" without careful analysis. That is inconsistent with the principle of seeking the truth from facts and garbles historical materials in order to prove one's own preconceptions. Admittedly, it is extremely important for a revolutionary historian "to have the courage to speak the truth." Yet if one deviates from seeking truth from facts and mistakes fabrication for the truth, the more "courage" one has, the more mistakes one will make.

Wang Zhizhong has gone to the opposite extreme and draws a direct connection between the Boxers and the Red Guards. This is made obvious in the last paragraph of his article in *Lishi yanjiu*: "It is proper to dispose of the new obscurantism and to make a clean break with the mistakes committed by myself and other young people of our generation more than ten years ago, when we were fooled by swindlers such as Lin Biao and the Gang of Four." That is why the author "emphasizes more the negative aspects of the Boxer movement" and satirizes or mocks the Boxers with words that bear a strong resemblance to those used to condemn the Great Proletarian Cultural Revolution.[39] For example: "Different stories were told by their contemporaries about the Boxers' discipline. As a rule those who advocated suppression would say that the Boxers beat, smashed, looted, burglarized, and killed" in their confrontations with the foreigners, while among the Chinese "they knocked down and swept away all new modes of production that were alien to feudalism."

Wang Zhizhong also magnifies the Boxers' weaknesses and shortcomings. He judges them to be "an inert force in history" that is "really hard to explain by means of the theory of mass movement that argues that such movements have principal and secondary aspects." I think highly of Wang Zhizhong's courage in facing squarely the question of success or failure in history, but the means by which Wang criticizes the Boxers is to make a forced analogy between the Red Guards and the Boxers.

The Boxers and the Red Guards are things of different natures that happened at different times and under different historical circumstances. They have neither internal connections nor historical links between them, and they should not be discussed at the same time. To prepare public opinion for their usurpation of party leadership and state power, Lin Biao and the Gang of Four deliberately compared the Boxers and the Red Guards. By using innuendoes in historical writings, they inordinately praised the Boxer movement in order to instigate further rebellious acts from the Red Guards. There is nothing unusual in all of this.

Today our task is to liquidate their pernicious influence so as to set to right what was thrown into disorder. We must repudiate not only the fallacies they spread, but also the method they used in historical studies. In his criticism of the feudalistic obscurantism and xenophobia of the Boxers, Wang Zhizhong attempted to repudiate the Red Guard movement. He believes that thereby he is rendering a service to the promotion of science and culture, as well as to the importation of advanced technology for the four modernizations. Still, for all his good intentions, Wang knowingly follows an incorrect path in his methodology, and this inevitably lead him into a series of errors in his arguments.

The most fundamental problem is that Wang Zhizhong has failed to adhere strictly to the basic tenets of Marxism. He quotes extensively from the Marxist classics in his analysis of problems, so that we know that he attaches great importance to Marxist theory. He writes: "My overall view is that we should bring to light and earnestly analyze the Boxers' backwardness and their feudal obscurantism, while at the same time affirming their anti-imperialism and just character. I agree that we must neither negate their historical role nor willfully exaggerate it. In short, we must seek truth from facts." These views, obviously, are consistent with the fundamental tenets of Marxism. Yet it is two different things to take Marxist theory seriously and to apply Marxist principles correctly. Unfortunately, it is in the application that Wang Zhizhong makes his mistakes. For example, his forced analogy between the Boxers and the Red Guards is a result of his deviation from the fundamental Marxist tenet of "making a concrete analysis of a concrete situation."

Departure from the principle that in Marxism particularity cannot be substituted for by generality causes Wang Zhizhong to apply mechanically to China the general principles that imperialism can "serve as the conscious tool of history." Marx put forth that analysis after investigating British rule in India, and he believed that "it was essential for those imported new modes of production to take root." It is Wang's deviation from the necessity of "looking at a problem from all sides" and his failure to identify the principal contradiction finally when determining the nature of a thing that makes him go astray. In appraising the Boxer movement, Wang negates it as "an inert force in history that is hard to explain by means of the theory of mass movements which have both a principal and secondary aspects," even while he "affirms its anti-imperialist and just character."

Although Wang Zhizhong declares that he "agrees neither to negate it [the Boxer movement] totally nor to exaggerate it willfully," he is unable to hold to his intention and consequently cannot help but contradict himself. His mistakes clearly show how important it is for historians to study the fundamental tenets of Marxism and correctly apply them as our guides in historical research. In this respect we all have much to learn. Let us encourage each other in our endeavors to achieve a better understanding from our research on the Boxer movement.

Notes

1. Karl Marx, The Eighteenth Brumaire of Louis Bonaparte (New York: International Publishers, 1963), p. 124.

2. Zhongguo shehui kexueyuan, jindaishi yanjiusuo, jindaishi ziliao bianjishe (Editorial Office for Modern History Materials, Institute of Modern History, Chinese Academy of Social Sciences), eds., Shandong yihetuan anjuan (Documents on the Shandong Boxers; hereafter YHTAJ) (Jinan: Qilu shushe, 1980), p. 45.

3. Sawara Tokusuke, "Quanluan jiwen" (Records of the Boxer Disturbances), in Qian Bocan et al., eds., Yihetuan (The Boxers, hereafter YHT) (Beijing: Shenzhou guoguangshe, 1951), 1:107.

4. Sawara Tokusuke, "Quanshi zaji" (Notes on the Boxer Affair), YHT 1:272.

5. Liu Yitong, "Minjiao xianchou dumen wenjianlu" (Observations on the Hostility between Our People and the Christians), in YHT 2:193.

6. Sawara, "Quanshi zaji," p. 246.

7. Memorial of the Censor Huang Guijun in Guojia danganju, Ming Qing danganguan (Ming-Qing Archive of the National Archives), eds., Yihetuan dangan shiliao (Archival Material on the Boxers; hereafter DASL) (Beijing: Zhonghua shuju, 1959), 1:45.

8. "Youguan yihetuan yulun" (Public Opinion on the Boxers), YHT 4:243.

9. DASL 1:162.

10. Mi Rucheng, Zhongguo jindai tielu shi ziliao (Materials on the History of Railroads in Modern China), 2:684.

11. Sawara, "Quanluan jiwen," p. 123.

12. Sawara, "Quanshi zaji," p. 246.

13. Sawara, "Quanluan jiwen," p. 162.

14. Chai E, "Gengxin jishi" (Records of 1900-1901), YHT 1:305-308.

15. "The Chinese Revolution and the Chinese Communist Party," Selected Works of Mao Tse-tung (Beijing: Foreign Languages Press, 1967), 2:309.

16. Collected Works of Marx and Engels (New York: International Publishers, 1967), 112:217.

17. "The Chinese Revolution and the Chinese Communist Party," Selected Works of Mao Tse-tung, 2:310.

18. Collected Works of Marx and Engels, 112:218.

19. "Wadexi quanluan biji" (Count von Waldersee's Diary on the Boxers), YHT 3:86.

20. "The Chinese Revolution and the Chinese Communist Party," p. 310.

21. Collected Works of Marx and Engels, 112:218.

22. Ibid., p. 416.

23. Ibid., p. 132.

24. Ibid., p. 220.

25. Ibid., p. 221.

26. "The Chinese War," Collected Works of Lenin (New York: International Publishers, 1929), 4:63.

27. Ibid., p. 59.

28. "The Bankruptcy of the Idealist Conception of History," Selected Works of Mao Tse-tung, 4:454.

29. Yihetuan yundong shiliao congbian (Collected Materials on the Boxer Movement) (Beijing: Zhonghua shuju, 1964), part 2, p. 47.

30. "Tianjin yiyue ji" (Record of a Month in Tianjin), YHT 2:157.

31. Zhongguo shehui kexueyuan, jindaishi yanjiusuo (Institute of Modern History, Chinese Academy of Social Sciences), eds., Gengzi jishi (Records of 1900) (Beijing: Zhonghua shuju, 1978).

32. "Gengzi shiguan beiwei ji" (The Siege of the Legations in 1900), YHT 2:358.

33. DASL 1:28.

34. DASL 1:49.

35. Zhongguo jindai duiwai guanxi shi (A History of China's Modern Foreign Relations), vol. 1, part 2, p. 141.

36. DASL 1:6.

37. Collected Works of Stalin (in Chinese), 4:125-26.

38. Frederick Engels, "Persia and China," On Colonialism (New York: International Publishers, 1972), p. 124.

39. Chai E, "Gengxin jishi," pp. 305-308.

40. Sun Zuomin, "Dui dangqian yihetuan yundong pingjia de jidian kanfa" (Some Thoughts on the Current Evaluations of the Boxer Movement), Dongyue luncong (Shandong Essays) 4 (1980).

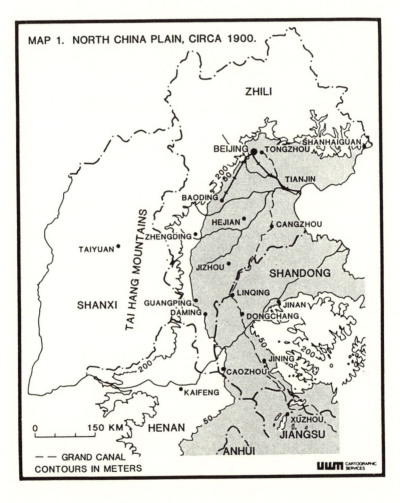

MAP 1. NORTH CHINA PLAIN, CIRCA 1900.

ZHILI

BEIJING TONGZHOU SHANHAIGUAN

TIANJIN

BAODING

HEJIAN CANGZHOU

ZHENGDING

TAIYUAN

JIZHOU

SHANDONG

LINQING

SHANXI GUANGPING JINAN

DAMING DONGCHANG

JINING

CAOZHOU

KAIFENG

0   150 KM   HENAN XUZHOU

JIANGSU

— — GRAND CANAL   ANHUI

CONTOURS IN METERS

TAI HANG MOUNTAINS

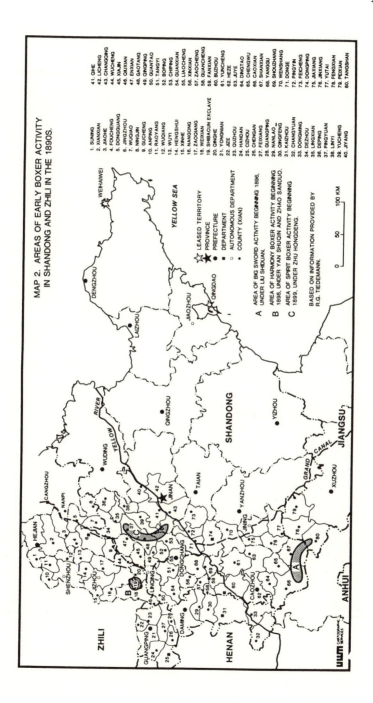

MAP 2. AREAS OF EARLY BOXER ACTIVITY IN SHANDONG AND ZHILI IN THE 1890S.

☆ LEASED TERRITORY
★ PROVINCE
■ PREFECTURE
● DEPARTMENT
○ AUTONOMOUS DEPARTMENT
○ COUNTY (XIAN)

A AREA OF BIG SWORD ACTIVITY BEGINNING 1896, UNDER LIU SHUAN.

B AREA OF HARMONY BOXER ACTIVITY BEGINNING 1898, UNDER YAN SHUQIN AND ZHAO SANDUO.

C AREA OF SPIRIT BOXER ACTIVITY BEGINNING 1899, UNDER ZHU HONGDENG.

BASED ON INFORMATION PROVIDED BY R.G. TIEDEMANN.

0   50   100 KM

1. SUNING
2. XIANXIAN
3. JIAOHE
4. FOUCHENG
5. DONGGUANG
6. JINGZHOU
7. WUQIAO
8. NINGJIN
9. GUCHENG
10. ANPING
11. RAOYANG
12. WUQIANG
13. WUYI
14. HENGSHUI
15. XINHE
16. NANGONG
17. ZAOQIANG
18. WEIXIAN
19. SHIBACUN EXCLAVE
20. QINGHE
21. YONGNIAN
22. JIZE
23. QUZHOU
24. HANDAN
25. CIZHOU
26. CHENGAN
27. FEIXIANG
28. GUANGPING
29. NANLAO
30. QINGFENG
31. KAIZHOU
32. CHANGYUAN
33. DONGMING
34. DEZHOU
35. LINGXIAN
36. DEPING
37. PINGYUAN
38. LINYI
39. YUCHENG
40. JIYANG
41. QIHE
42. LICHENG
43. CHANGQING
44. WUCHENG
45. XIAJIN
46. QIUXIAN
47. ENXIAN
48. GAOTANG
49. QINGPING
50. GUANTAO
51. TANGYI
52. BOPING
53. CHIPING
54. GUANXIAN
55. LIAOCHENG
56. XINXIAN
57. ZAOCHENG
58. GUANCHENG
59. FANXIAN
60. BUZHOU
61. YUNCHENG
62. HEZE
63. JUYE
64. DINGTAO
65. CHENGWU
66. CAOXIAN
67. SHANXIAN
68. YANGGU
69. SHOUZHANG
70. WENSHANG
71. DONGE
72. PINGYIN
73. FEICHENG
74. DONGPING
75. JIAXIANG
76. JINXIANG
77. YUTAI
78. FENGXIAN
79. PEIXIAN
80. TANGSHAN

ZHILI

HEJIAN

SHENZHOU

ZEZHOU

GUANGPING

DAMING

HENAN

CANGZHOU

NANPI

WUDING

YELLOW RIVER

QINGZHOU

JINAN

TAIAN

DONGCHANG

LINQING

CAOZHOU

SHANDONG

YANZHOU

JINING

GRAND CANAL

YIZHOU

XUZHOU

JIANGSU

ANHUI

DENGZHOU

WEHAIWEI

LAIZHOU

QINGDAO

JIAOZHOU

YELLOW SEA

UWM CARTOGRAPHIC SERVICES

222

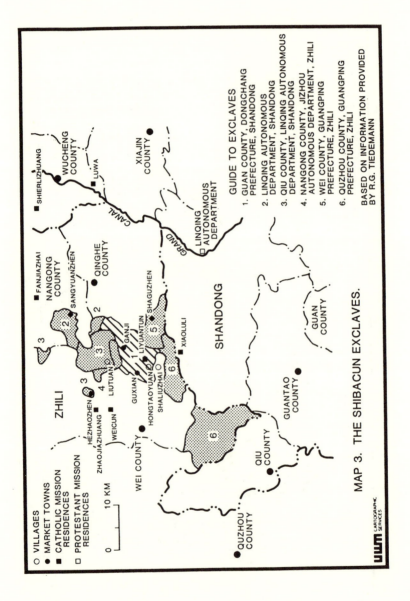

MAP 3. THE SHIBACUN EXCLAVES.

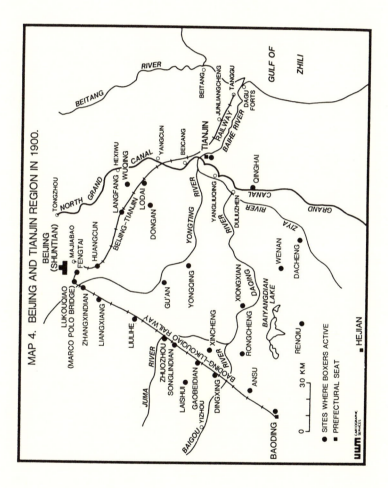

MAP 4. BEIJING AND TIANJIN REGION IN 1900.

GULF OF ZHILI

BEITANG RIVER
BEITANG
JUNLIANGCHENG
TANGGU
DAGU FORTS
RAILWAY
BAIHE RIVER
TIANJIN

TONGZHOU
NORTH GRAND CANAL
HEXIWU
LANGFANG
WUQING
YANGCUN
BECANG

BEIJING (SHUNTIAN)
MAJIABAO
FENGTAI
HUANGCUN
BEIJING-TIANJIN
LODAI
DONGAN
YONGTING RIVER
YANGLIUQING
QINGHAI
GRAND CANAL
ZIYA RIVER

LUKOUQIAO (MARCO POLO BRIDGE)
ZHANGXINDIAN
LIANGXIANG
LIULIHE
GU'AN
YONGQING
XINCHENG
XIONGXIAN
DAQING
DULIUZHEN
WENAN
DACHENG

JUMA RIVER
ZHUOZHOU
SONGLINDIAN
BAODING-LUKOUQIAO RAILWAY
 BAODING-LIULIHE RIVER
RONGCHENG
BAIYANGDIAN LAKE
RENQIU

LAISHUI
GAOBEIDIAN
DINGXING RIVER
ANSU

BAIGOU
YIZHOU

BAODING

HEJIAN

• SITES WHERE BOXERS ACTIVE
■ PREFECTURAL SEAT

0   30 KM

UWM CARTOGRAPHIC SERVICES

Glossary to

Recent Studies of the Boxer Movement

Edited by David D. Buck

M. E. Sharpe, Inc.
80 Business Park Drive, Armonk, NY 10504

Glossary

A

安次（县）	Anci	安陵集	Anlingji
安平（县）	Anping	安肃（县）	Ansu

B

霸（县）	Ba	八卦	bagua
八卦教	baguajiao	白莲教	bailianjiao
拜上帝会	baishangdi hui	拜团	baituan
白阳教	baiyangjiao	保定（府）	Baoding
保甲局	baojiaju	霸州	Bazhou
北台吉	Beitaiji	本明	Benming
滨州（县）	Binzhou	博平（县）	Boping

C

沧州（府）	Cangzhou	曹（县）	Cao
曹得礼	Cao Deli	曹福田	Cao Futian
曹倜	Cao Ti	曹振镛	Cao Zhenyong
曹州（府）	Caozhou	查荣绥	Cha Rongsui
长乐（县）	Changle	长清（县）	Changqing
长阳（县）	Changyang	陈胜	Chen Sheng
陈廷经	Chen Tingjing	程明洲	Cheng Mingzhou
城武（县）	Chengwu	陈庄	Chenzhuang
茌平（县）	Chiping	《茌平县志》	Chiping xianzhi
《崇陵传信录》	Chongling chuan xin lu	川心	Chuanxin
《出劫纪略》	Chujie jilue	传神 助教 灭洋 共和义和团	Chuan shen, zhu jiao' mie yang, gong he yi he tuan
传帖相约	Chuantie xiangyue	慈喜	Cixi
崔士俊	Cui Shijun		

D

大乘	dacheng		
大刀会	dadaohui	大佛寺	Dafosi
大沽	Dagu	大贵	Dagui
大红拳	dahongquan	戴玄之	Dai Xuanzhi

· 1 ·

打灭洋教 重兴 damie yangjiao   大名（县）   Daming
拳会 chongxing quanhui

砀山（县） Dangshan   刀枪不入 daoqiangburu
道通 Daotong   大师兄 dashixiong
大兴（县） Daxing   大芝坊村 Dazhifangcun
大足（县） Dazu   得胜饼 deshengbing
德州（县） Dezhou   丁名楠 Ding Mingnan
丁棕匠 Ding Zongjiang   定陶（县） Dingtao
定兴（县） Dingxing   抵制洋人洋教 dizhi yangren yangjiao

董福祥 Dong Fuxiang   东昌（府） Dongchang
栋臣 Dongchen   东光（县） Dongguang
东湍 Dongtuan

E

恩（县） En
二狼拳 erlangquan   二毛子 ermaozi
二圣堂村 Ershengtangcun   二师兄 ershixiong

F

范文澜 Fan Wenlan   方荣升 Fang Rongsheng
方诗铭 Fang Shiming   飞地 feidi
冯克善 Feng Keshan   《封神榜》 Fengshenbang
丰台 Fengtai   抚 fu

G

干集 Ganji   高小麻子 Gao Xiaomazi
高碑店 Gaobeidian   高唐琉璃寺 Gaotang Liuli si
《庚辛纪事》 Gengxin jishi   《庚子奉禁义和 Gengzi fengjin Yihe
 拳汇录》 quan huilu

《庚子记事》 Gengzi jishi   《庚子莘蜂录》 Gengzi pingfeng lu
《庚子西狩丛谈》 Gengzi xishou congtan   《庚子西行记事》 Gengzi xixing jishi
《庚子义和团运 Gengzi yihetuan yundong shimo
动始末》

功夫 gongfu   冠（县） Guan
管鹤 Guan He   官兵 guanbing

· 2 ·

官督民办	guandu minban	关公	Guangong
《冠县志》	Guan xianzhi	《光绪庚子我的金钟罩团的组织经过》	Guangxu Gengzi wo de jinzhongzhao tuan de zuzhijingguo
《广益丛报》	Guangyi congbao	广宗（县）	Guangzong
馆陶（县）	Guantao	故城（县）	Gucheng
《古春草堂笔记》	Guchuncaotang biji	崮堆	Gudui
鼓浪屿	Gulangyu	郭栋臣	Guo Dongchen
《郭栋臣的亲笔回忆》	Guo Dongchen de qinbi huiyi		
郭沫若	Guo Moruo	《国闻报》	Guowenbao

H

海丰（县）	Haifeng	海门（县）	Haimen
韩二	Han Er	韩以礼	Han Yili
哈尔滨	Harbin	何建德	He Jiande
何乃莹	He Naiying	何如道	He Rudao
河间（府）	Hejian	黑龙江	Heilongjiang
洪钧	Hong Jun	洪用舟	Hong Yongzhou
红灯罩	hongdengzhao	红门	hongmen
红拳	hongquan	红砖会	hongzhuanhui
后李庄	Houlizhuang	侯魏村	Houweicun
胡宾	Hu Bin	胡绳	Hu Sheng
滑（县）	Hua	黄当选	Huang Dangxuan
黄育梗	Huang Yubian	会	hui
《汇报》	Hui bao	惠民（县）	Huimin
虎神营	Hushenying	虎尾鞭	huweibian

J

翦伯赞	Jian Bozan	蒋楷	Jiang Kai
江油（县）	Jiangyou	姜子牙	Jiang Ziya
剿	jiao	教	jiao
《教案奏议汇编》	Jiaoan zouyi huibian	《教会源流考》	Jiaohui yuanliu kao
教主	jiaozhu	吉林	Jilin
金冲及	Jin Chongji	金家瑞	Jin Jiarui
景廷宾	Jing Tingbin	经额布	Jingebu
静海（县）	Jinghai	景善	Jingshan
《景善日记》	Jingshan riji	敬神附体	jingshen futi

罗荣光	Luo Rongguang	落垡	Luofa
罗教	Luojiao	《驴背集》	Lübeiji

M

马金叙	Ma Jinxu	马玉昆	Ma Yukun
茂州（州）	Maozhou	梅花拳	meihuaquan
秘密结社	mimijieshe	秘密宗教，教门	mimi zongjiao, jiaomen
明教寺	Mingjiaosi		
《明清之际宝卷文学与白莲教》	Ming Qing zhiji zhi baojuan wenxue yu bailianjiao	明兴	Mingxing
民团	mintuan	密云（县）	Miyun

N

纳继成	Na Jicheng	纳尔经额	Naerjinge
南宫（县）	Nangong	南乐（县）	Nanle
《南苑三台山神字碑》	Nanyuan santaishan shenzibei		
那桐	Natong	那彦成	Nayancheng
哪吒	Nazha	《聂将军歌》	Nie Jiangjun ge
聂士成	Nie Shicheng	牛庄	Niuzhuang

P

庞三杰	Pang Sanjie	庞维	Pang Wei
庞家林庄	Pangjialinzhuang	潘庄	Panzhuang
裴秀亭	Pei Xiuting	配义团	Peiyituan
彭家屏	Pengjiaping	平原（县）	Pingyuan
《平原拳匪记事》	Pingyuan quanfei jishi	《破邪详辩》	Poxie xiangbian
溥静	Pujing	浦台	Putai

Q

戚其章	Qi Qizhang	乾卦	qiangua
齐河（县）	Qihe	清（县）	Qing
《清端郡王载漪封爵后杂记梗概》	Qing Duanjun wang Zayi fengjuehou zaji genggai	清茶门教	Qingchamenjiao
		清丰（县）	Qingfeng
		《清高宗实录》	Qing gaozong shilu
清河（县）	Qinghe	庆溥	Qingpu
清水（县）	Qingshui	庆云（县）	Qingyun
秦力山	Qin Lishan	琦善	Qishan
《邱莘教匪》	Qiuxin jiaofei	渠（县）	Qu
《拳案杂存》	Quanan zacun	拳匪	quanfei
《拳匪祸国记》	Quanfei huoguoji	《拳匪记略》	Quanfei jilüe

孙祚民	Sun Zuomin	孙逸仙	Sun Yat—sen
		T	
太平（县）	Taiping	坛	tan
檀玑	Tan Ji	唐恒乐	Tang Huanle
《唐氏三先生集》	Tang Shisan xiansheng ji	堂邑（县） 坛口镇	Tangyi Tankouzhen
陶成章	Tao Chengzhang	天地会	tiandihui
		《天津一月记》	Tianjin yiyueji
天理会	tianlihui	天龙	tianlong
《天龙八部	Tianlong babu	天龙八卦教	Tianlong baguajiao
元明册》	yuanmingce	天门	tianmen
铁布衫	tiebushan	廷杰	Ting Jie
廷雍	Tingyong	童振青	Tong Zhenqing
同盟会	tongmenghui	同仁	tongren
同兴	Tongxing	通州（州）	Tongzhou
秃闺女	Tu Guinu	团练	tuanlian
团民	tuanmin	驼儿山	Tuoershan
		W	
碗儿庄	Wanerzhuang	王秉衡	Wang Bingheng
王成德	Wang Chengde	王敬修	Wang Jingxiu
王立言	Wang Liyan	王伦	Wang Lun
王普仁	Wang Puren	王庆一	Wang Qingyi
王致中	Wang Zhizhong	《万国公报》	Wanguo gongbao
威（县）	Wei	魏村	Weicun
《威县志》	Weixian zhi	文	wen
文顺	Wenshun	文琭	Wenli
武	wu	吴广	Wu Guang
吴楷	Wu Kai	吴官屯	Wuguantun
吴桥（县）	Wuqiao	武清（县）	Wuqing
武圣关帝	Wusheng guandi	晤修	Wuxiu
武邑（县）	Wuyi		
		X	
隰（县）	Xi	夏津（县）	Xiajin
项得胜	Xiang Desheng	香山	Xiangshan
乡团	xiangtuan	孝义（县）	Xiaoyi
邪道	xiedao	协和	xiehe
邪会	xiehui	邪教	xiejiao
莘（县）	Xin	辛店	Xindian

兴清灭洋	XingQing mieyang	兴中灭洋	XingZhong mieyang
		《新闻报》	Xinwenbao
《信阳县志》	Xinyang xianzhi	雄（县）	Xiong
《西游记》	Xiyouji	西朱管营	Xizhuguanying
徐安帼	Xu Anguo	许景澄	Xu Jingcheng
徐福	Xufu		

Y

阎书勤	Yan Shuqin	《郯城县具沂	Yanchengxian ju
杨福同	Yang Futong	州府禀》	Yizhoufu bing
杨慕时	Yang Mushi		
杨寿臣	Yang Shouchen	杨四海	Yang Sihai
阳谷（县）	Yanggu	洋民	Yangmin
阳信（县）	Yangxin	《燕京学报》	Yanjing xuebao
盐山（县）	Yanshan	兖州（府）	Yanzhou
姚洛奇	Yao Luoqi	义合	yihe
义和会	yihehui	义和拳	yihe quan
《义和团》	Yihetuan	《义和团档案史料》	Yihetuan dangan shiliao
《义和团的前史》	Yihetuan de qianshi	《义和拳教门源流考》	Yihequan jiaomen yuanliukao
《义和团事实》	Yihetuan shishi		
《义和团研究》	Yihetuan yanjiu	义和团勇	Yihe tuanyong
《义和团源流试探》	Yihetuan yuanliu shitan	《义和团运动》	Yihetuanyundong
		《义和团运动初期斗争阶段的几个问题》	Yihe tuan yundong chuqi douzheng jie duan de jige wenti
《义和团的兴起和失败》	Yihetuan de xing-qi he shibai	《义和团运动史料丛编》	Yihetuan yundongshi liao congbian
义和乡团	yihexiangtuan		
义和 邪教	yihexiejiao	异伙拳	yihuoquan
奕劻	Yikuang	义民	yimin
郾（县）	Ying	英和	yinghe
英年	Yingnian	印务	Yinwu
《郾县志》	Yinxianzhi	沂州（府）	Yizhou
亦准	Yizhun	玉文义	Yu Wenyi
《袁李碑文》	Yuanli beiwen	袁世敦	Yuan Shidun
袁世凯	Yuan Shikai	袁 昶	Yuan Chang
元城（县）	Yuancheng	裕录	Yulu
恽毓鼎	Yun Yuding	毓贤	Yuxian

Z

载澜	Zailan	载濂	Zailian
载勋	Zaixun	载漪	Zaiyi
造反	zaofan	枣强（县）	Zaoqiang
增祺	Zeng Qi	张天师	Zhang Tianshi
张德成	Zhang Decheng	张 衡	Zhang Heng
张景文	Zhang Jingwen	张联恩	Zhang Lianen
张 嶙	Zhang Lin	张洛焦	Ziang Luojiao
张汝梅	Zhan Rumei	张士杰	Zhang Shijie
张之洞	Zhang Zhidong	张登（县）	Zhangdeng
张官寨	Zhangguanzhai	章丘（县）	Zhangqiu
沽化（县）	Zhanhua	赵金环	Zhao Jinhuan
赵老祝	Zhao Laozhu	赵老广	Zhao Laoguang
赵三多	Zhao Sanduo	赵舒翘	Zhao Shuqiao
赵家庄	Zhaojiazhuang	赵庄	Zhaozhuang
震（卦）	zhen	郑炳麟	Zheng Binglin
郑元善	Zheng Yuanshan	正定（县）	Zhengding
真空家乡	zhenkong jiaxiang	真空家乡	zhenkong jiaxiang
		无生父母	wusheng fumu
无身父母	wushen fumu	知道了	zhidaole
《致胡绍箴书》	Zhi Hu Shaojian shu	《直东剿匪电存》	Zhidong jiaofei diancun
		《中国近代反帝反	Zhongguojindai fandi fan-
志和团	zhihetuan	封建历史歌谣选》	fengjian lishi geyao xuan
《中国近代史》	Zhongguo jindaishi		
《中国近代史	Zhongguo jindai		
资料》	shiz ilao	《中国史纲要》	Zhongguo shigangyao
		《中国史稿》	Zhongguo shigao
《中国小说史略》	Zhongguo xiaoshuo shilue	忠义军	zhongyijun
		朱红灯	Zhu Hongdeng
朱九斌	Zhu Jiubin	朱 勋	Zhu Xun
朱祖谋	Zhu Zumou	庄亲王载勋	Zhuang Qinwangzaixun
诸葛亮	Zhuge Liang	朱家寨	Zhujiazhai
涿州（府）	Zhuozhou	助清灭洋	zhuQing mieyang
资阳（县）	Ziyang	总管大师兄	zongguan dashixiong
总理衙门	Zongli yamen	邹平（县）	Zouping
遵化（县）	Zunhua	捉拿洋教 振兴	zuona yangjiao Zhenxing
左云（县）	Zuoyun	中国	Zhongguo
祖 师	zushi	祖师爷	zushiye